I0478611

US Navy Fleet Submarine Manual

NavPers 16160

Produced for ComSubLant by
Standards and Curriculum Division
Training, Bureau of Naval Personnel

June 1946

RESTRICTED

CONTENTS

SUBMARINE OPERATIONS

SUBMARINE TRAINING

1
DEVELOPMENT OF THE SUBMARINE

A. EARLY UNDERWATER DEVICES

1A1. Early Greek devices. The submarine first became a major factor in naval warfare during World War I, when Germany demonstrated its full potentialities. However, its advent at that time, marked by wholesale sinkings of Allied shipping, was in reality the culmination of a long process of development.

Ancient history includes occasional records of attempts at underwater operations in warfare. The Athenians are said to have used divers to clear the entrance of the harbor of Syracuse during the siege of that city; during his operations against Tyre, Alexander the Great ordered divers to impede or destroy any submarine defenses the city might undertake to build. But in none of these records is there a direct reference to the use of submersible apparatus of any kind. There is, however, a legend that Alexander the Great himself made a descent into the sea in a device which kept its occupants dry and admitted light.

In the Middle Ages, the Arabian historian Boha-Eddin reports that a diver using submersible apparatus succeeded in gaining entrance into Ptolemais (Acre) during the siege of that city in A.D. 1150. And in 1538, a diving bell was built and tested at Toledo, Spain. Although it attracted the attention of the Emperor Charles V, the device was never further developed and passed quickly into oblivion.

1A2. Bourne's idea. Not until 1580 sponsored by one Magnus Pegelius was launched in 1605. But the designers made one serious oversight. They failed to consider the tenacity of underwater mud, and the craft was buried at the bottom of a river during initial underwater trials.

1A3. Van Drebel's submersible. It is to Cornelius Van Drebel, a Dutch physician, that credit is usually given for building the first submarine. To him is conceded the honor of successfully maneuvering his craft, during repeated trials in the Thames River, at depths of 12 and 15 feet beneath the surface.

Van Drebel's craft resembled those of Bourne and Pegelius in that its outer hull consisted of greased leather over a wooden framework. Oars, extending through the sides and sealed with tight-fitting leather flaps, provided propulsion either on the surface or when submerged. Van Drebel built his first boat in 1620 and followed it later with two others both larger but embodying the same principles. It is reported that after repeated tests, James I took a trip in one of the larger models and demonstrated its safety. But despite this evidence of royal favor, the craft failed to arouse the interest of the navy in an age when all conception of the possibilities of submarine warfare was still far in the future.

1A4. Eighteenth century plans. Submarine boats seem to have been numerous in the early years of the 18th century. By 1727 no fewer than 14 types

does any record appear of a craft designed to be navigated under water. In that year, William Bourne, a British naval officer, made designs of a completely enclosed boat which could be submerged and rowed under the surface. The device consisted of a wooden framework covered with waterproofed leather. It was to be submerged by reducing its volume as a result of contracting the sides through the use of hand vises. Although Bourne never built this boat, a similar construction

had been patented in England alone.

An unidentified inventor whose work is described in the *Gentleman's Magazine* for 1747 introduced an ingenious device for submerging and surfacing his submarine. His craft was to have had a number of goatskins built into the hull, each of which was to be connected to an aperture in the bottom. He planned to submerge the vessel by filling the skins with water, and to bring it to the

surface again by forcing the water out of the skins with a "twisting rod." This seems to have been the first approach to the modern ballast tank. By that time, ideas were plentiful, some of them fanciful and grotesque, but some containing elements capable of practical

application. Lack of full understanding of the physical and mechanical principles involved, coupled with the well-nigh universal conviction that underwater navigation was impossible and of no practical value, postponed for more than another hundred years the attempt to utilize a submarine in warfare.

B. EARLY SUBMARINES

1B1. David Bushnell's *Turtle*. During the American Revolutionary War, a submarine was first used as an offensive weapon in naval warfare. The *Turtle*, a one-man submersible invented by David Bushnell and hand-operated by a screw propeller, attempted to sink a British man-of-war in New York Harbor. The plan was to attach a charge of gunpowder to the ship's bottom

with screws and explode it with a time fuse. After repeated failures to force the screws through the copper sheathing on the hull of the HMS *Eagle,* the submarine gave up, released the charge, and withdrew. The powder exploded without result, except that the *Eagle* at once decided to shift to a berth farther out to sea.

1B2. Fulton's *Nautilus*. Although his name is most often associated with the invention of the steamboat, Robert Fulton experimented with submarines at least a decade before he sailed the *Clermont* up the Hudson. His *Nautilus*

development of the craft, even though his model displayed some of the best features of any submarine up to that time.

1B3, The Confederate "Davids". Development of the submarine boat was

was built of steel in the shape of an elongated oval, and was somewhat similar in structure to today's submarine. A sail was employed for surface propulsion and a hand-driven propeller drove the boat

when submerged. A modified form of conning tower was equipped with a porthole for observation, since the periscope had not yet been invented. In 1801, Fulton tried to interest France, Britain, and America in his idea, but no nation ventured to sponsor the

held back during all of this period by lack of any adequate means of propulsion. Nevertheless, inventors continued resolutely with

experiments upon small, hand-propelled submersibles carrying a crew of not more than six or eight men. On 17 February 1864, a Confederate vessel of this type sank a Federal corvette that was blocking Charleston harbor. This first recorded instance of a

3

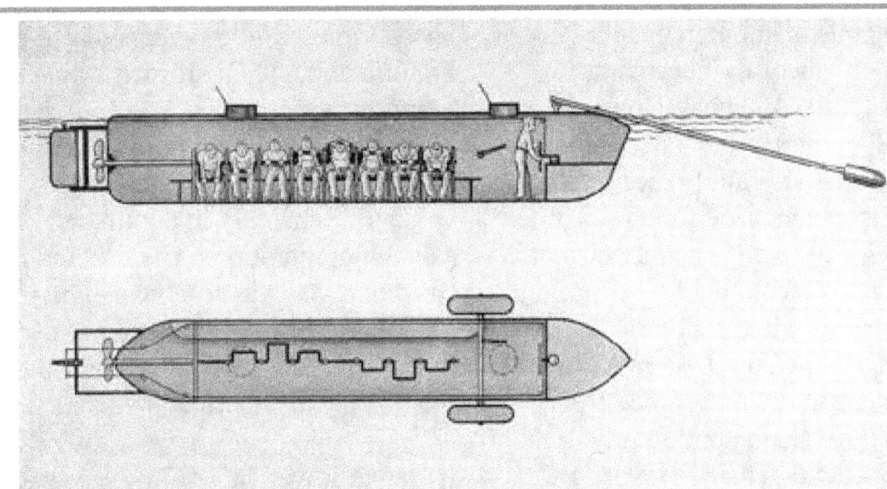

Figure 1-3. The HUNTLEY, one of numerous "Davids" constructed during the War Between the States.

submarine sinking a warship was accomplished by a torpedo suspended ahead of the bow of the *Huntley* as she rammed the *Housatonic*.

1B4. Garrett's steam propulsion. Interest in the improvement of the submarine was active during the period of the War Between the States, but the problem of a suitable means of propulsion continued to limit progress. Steam was tried and finally in 1880 an English clergyman, the Rev. Mr. Garrett, successfully operated a submarine with steam from a coal-fired

could submerge to a depth of 50 feet, was fitted with one of the first practical torpedo tubes.

1B5. Electric propulsion. Meanwhile, electric propulsion machinery had proved its utility in many fields, and in 1886, an all-electric submarine was built by two Englishmen, Campbell and Ash. Their boat was propelled at a surface speed of 6 knots by two 50-horsepower electric motors operated from a 100-cell storage battery. However, this craft suffered one major handicap; its batteries had to be recharged and overhauled at such short

boiler which featured a retractable smokestack. During the same period, a Swedish gun designer, Nordenfelt, also constructed a submarine using steam and driven by twin screws. His craft, which

intervals that its effective range never exceeded 80 miles.

C. MODERN SUBMARINES

1C1. Holland's *Plunger*. Antedating the efforts of Nordenfelt were the experiments of J. P. Holland of New Jersey, who launched his first boat in 1875. Although his early models embodied features that were discontinued as development progressed, many of his initial ideas, perfected in practice, are in use today. Outstanding in importance was

the principle of submergence by water ballast, and the use of horizontal rudders to dive the boat. However, not until 1895, did Holland, in competition with Nordenfelt, finally receive an order for a submarine from the United States Government. The vessel was propelled by steam on the surface and by electricity when submerged. This craft was named the *Plunger*. The original craft was redesigned frequently during construction and finally abandoned altogether in favor of a newer model already building in the Holland shipyard. This was Holland's ninth submersible, but it was the first to be delivered to the United States Government. It was delivered in 1900, and was the basic design of all British submarines to follow.

1C2. Lake's Submarines. Simon Lake, who began building submarines in 1894, designed them primarily with peacetime uses in mind.

His vessels could travel about on the sea bottom, and had an air lock which permitted a passenger in a diving helmet to emerge from the hull to walk about and explore. In fact, Lake used his vessels extensively in commercial salvaging operations. His first model, the *Argonaut, Jr.* was solely an experimental one. It was built of two layers of yellow pine with a sheet of canvas between them, and was operated by hand.

It was followed in 1897 by the *Argonaut*, a cigar-shaped hull 36 feet long and powered by a 30-horsepower gasoline engine. This craft could submerge to the bottom of a lake or river and roll along at bottom on three wheels; or, for navigating. The wheels could be raised and carried in packets in the keel. In 1898 the *Argonaut* traveled under its own power through heavy November storms from Norfolk to New York, and was thus the first

Figure 1-4. The ARGONAUT JR.

5

submarine to navigate extensively in the open sea. In 1906 Lake built the *Protector* and sold it to Russia. After it had successfully passed various severe tests there, Lake built a number of submersibles on contract for the Russian Government.

1C3. Conclusion. Thus the fundamental principles of construction and operation of submarine boats had been determined and demonstrated before the outbreak of World War I. By that time, too, internal combustion engines, both gasoline and Diesel, were available for use as practical power plants.

The invention of the periscope had materially increased the practical feasibility of underwater navigation. And the primary weapon of the submarine, the torpedo, had been perfected for use. Thus, the preliminary development of the submarine was finished, and the vessel was ready to take its place as a major factor in naval strategy. In place of the tiny, one-man contraptions that first dared to venture beneath the surface had come effective weapons, only a little short of the powerful, 70-man, fleet-type submarines that range the seas today.

D. GENERAL DATA

1D1. Type of Construction. When the submarine rests on the surface, so little of it is seen above the water that it has the appearance of being longer and more slender than

it really is. Actually, the modern fleet type submarine is approximately 312 foot long with a superstructure deck tapering almost to a point, both fore and aft, from its greatest width of approximately 16 feet amidship.

Figure 1-5. USS O-7 (1918)

6

Figure 1-6. USS R-6 (1919).

Figure 1-7. USS S-17 (1921)

Figure 1-8. USS S-46 (1925).

Figure 1-9. USS NARWHAL (1930).

8

Figure 1-10. USS BLACKFIN (1944).

Beneath the superstructure deck is the all-welded hull; actually it is two hulls, for the fleet-type submarine is a double-hull vessel. To understand the construction of a submarine, one must first appreciate the conditions under which the vessel operates below the surface. This means that the submarine must at all times be watertight, pressure. The fabrication of these containers into the hull of the vessel is illustrated in Figure 1-11.

Pressure vessels, while capable of withstanding great pressure, do not in themselves possess great rigidity. Being subject to mechanical action (leverage), they must be secured to each other by

otherwise self-destruction would result. The construction of the submarine, therefore, is on the basis of the fabrication of a series of watertight containers into one large watertight cylinder by means of watertight joints. However, since the submarine must operate at times at great depths, these watertight containers must be strong enough to withstand the pressure head of sea water at that depth. Therefore, the watertight containers must be *pressure vessels,* that is, watertight containers or cylinders capable of withstanding great

one common strength member (the keel), as well as by watertight connections (bulkheads). The submarine with its keel, pressure hull, and watertight bulkheads is shown in Figure 1-12.

In the double-hull type of submarine, the pressure hull is inside the outer hull; between the two hulls are the water and the fuel oil tanks. The double-hull construction extends from the after bulkhead of the forward torpedo room to the forward bulkhead of the after torpedo room. The pressure hull,

9

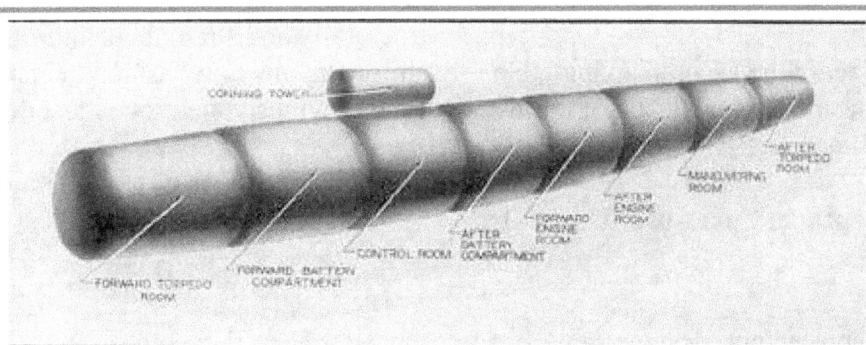

Figure 1-11. Type of construction, showing arrangement of compartments, without the superstructure or tanks.

or inner hull, extends from the forward bulkhead of the forward trim tank to the after bulkhead of the after trim tank. Above the hull is built a non-watertight superstructure which forms the main deck, for use when surfaced.

A gun, usually a 5"/25, wet type, is mounted topside. The space below the deck is used as locker space for stowing anchor gear, lines, and other gear that cannot be damaged by water. Ready ammunition in boxes and the ship's boat are also kept here.

The deck is perforated on either side with circular holes among the entire length to prevent air pockets from forming within the superstructure when it becomes flooded. A watertight tower, know as the *conning tower,* extends upward through the superstructure amidships. The top of the conning tower is used as a bridge when on the surface, but when submerged, the control of the boat is maintained either from the conning tower or from a compartment directly below it, known as the *control room.* Periscopes

Figure 1-12. Type of construction, showing the general arrangement of the superstructure.

operated from the conning tower extend above the bridge and are used for making observations when submerged.

1D2. Size. In the accompanying table are shown dimensional data of the fleet type submarine.

1D3. Depth and pressure. The modern fleet

Individual compartments are air tested for tightness only to a pressure of 15 psi. Watertight bulkheads are designed structurally and strengthened through reinforcements to withstand the pressure at the previously mentioned test depths.

The total pressure that the hull must

Displacement (designed)	1,523 tons
Displacement (surface)	1,816 tons (diving trim
Length (over-all)	311'-9"
Breadth (extreme)	27"-4"
Mean draft (surface)	15'-3" (diving trim)
Number of frames	139
Frame spacing (except 35 to 62 and 69 to 105 spacing 30")	24" center to center
Freeboard at stern	3'-11"
Freeboard at bow	12'-5"
Diameter pressure hull (max.)	16'-0 3/8"
Distance from keel to centerline of hull	12'-0"
Floodable space	
Forward torpedo room	4,481 cu. ft.
Forward battery compartment	4,056 cu. ft.

Control room	4,653 cu. ft.
Conning tower	760 cu. ft.
After battery compartment	5,821 cu. ft.
Forward engine room	4,535 cu. ft.
After engine room	4,277 cu. ft.
Maneuvering room	3,410 cu. ft.
After torpedo room	3,455 cu. ft.
Total floodable space	35,448 cu. ft.

Type submarine is built to withstand the pressure of a head of sea water, consistent with requirements as shown by battle experience and with the Bureau of Ships specifications. The pressure is measured in actual submergence tests from the surface of the water to the axis of the vessel through its pressure hull.

withstand is actually the differential pressure between the interior hull pressure and the external head of water at a given depth.

1D4. Main propulsion, speed and cruising radius. The average fleet type submarine is driven by four main propulsion diesel engines, each capable of producing 1600 hp.

11

Figure 1-13. Torpedo tubes.

Figure 1-14. Deck gun, 5"/25.

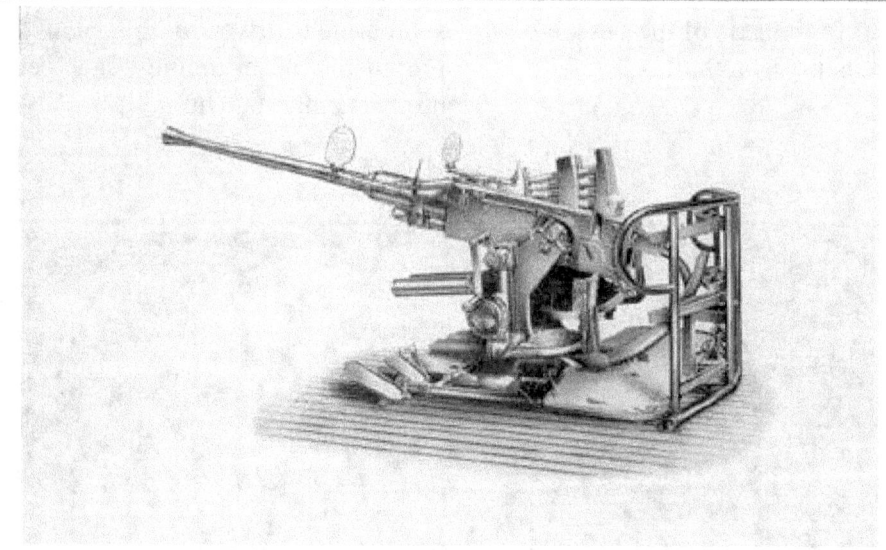

Figure 1-15. Antiaircraft gun, 40 mm.

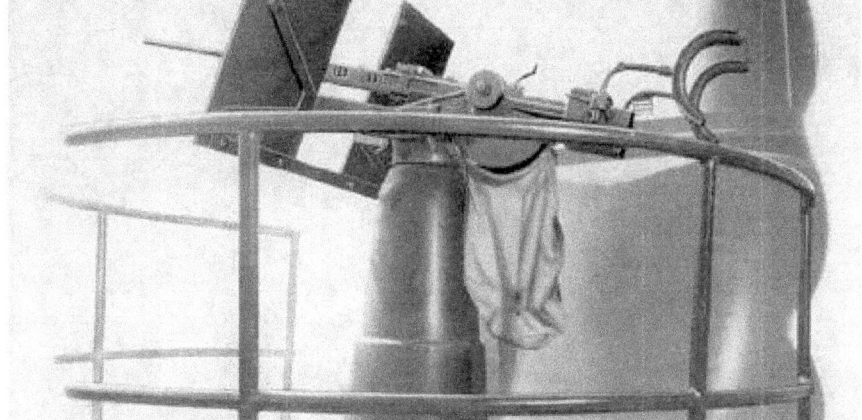

Figure 1-16. Antiaircraft gun, 20 mm.

13

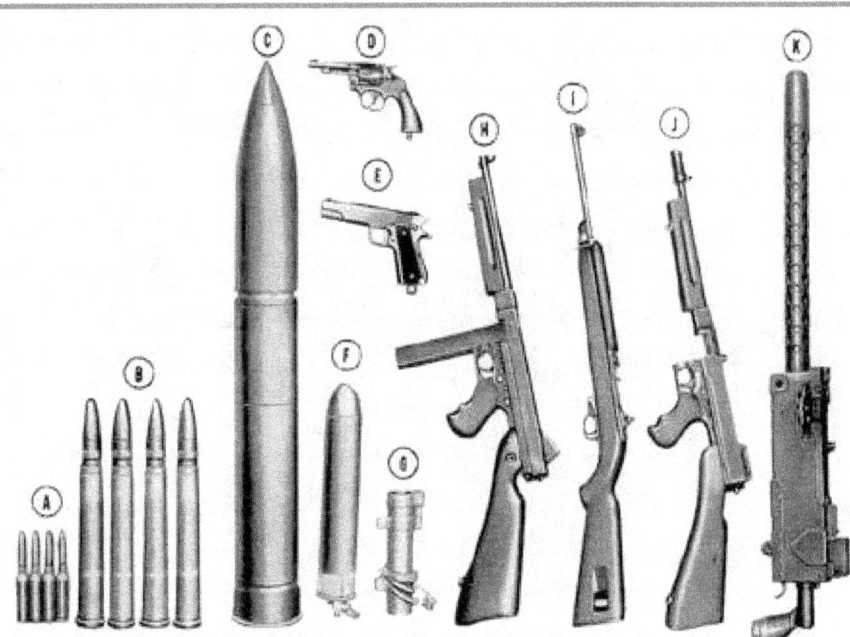

Figure 1-17. Small arms and pyrotechnics.

Figure 1-18. Conning tower, looking aft.

14

Figure 1-19. Lower portion of a modern periscope.

15

The four main generators each produce 1100 kw. There are four main motors, driven by the generators or batteries, The rollers aft of the tubes and the racks farther aft (not shown in the illustration) are used for torpedo reload. The tubes

each producing about 1375 hp. The two reduction gears are of the herringbone, 2-pinion type and produce about 2750 hp at each shaft. The auxiliary engine is rated at 450 hp and drives a 300-kw generator.

The average fleet type submarine is capable of a speed of about 21 knots when operating on the surface and approximately 10 knots when submerged. This submarine has a cruising range in excess of 12,000 miles.

1D5. Ship's complement and ship's armament. a. *Ship's complement.* The personnel aboard the fleet type submarine range in number from 66 to 78. Officers number from 6 to 8, and men from 60 to 70.

b. *Ship's armament. Torpedo tubes* (figure 1-13) are the main offensive and defensive armament of the submarine. A total of 10 21-in. tubes are carried, six forward and four aft. Those shown in the illustration are No. 1 and No. 2. Located in pairs below tubes No. 1 and No. 2 are Nos. 3 and 4, and Nos. 5 and 6. The upper half of the No. 3 tube is visible in the lower right corner of the illustration.

Immediately above tubes No. 1 and No. 2 is the torpedo tube blow and vent manifold used for blowing or venting the tubes, WRT tank, and the trim tank. (See Figure 1-13.)

can be fired electrically or by hand when surfaced or submerged. The condition of the tube is indicated by the torpedo ready lights, shown to the left of the No. 2 tube.

The *5"/25 deck gun* (Figure 1-14) is a dual purpose gun. It is so mounted as to be used effectively against surface craft and aircraft. Two guns may be carried: if one gun is carried it is located abaft of the conning tower.

The *40-mm antiaircraft gun* (Bofors), shown in Figure 1-15, is mounted forward of the conning tower. It is principally an antiaircraft weapon, but may be used against surface craft. It is a rapid fire, recoil type of gun. In some instances it is being replaced by a 37-mm gun. The 40-mm gun sometimes replaces two 20-mm guns.

The *20-mm antiaircraft gun* (Figure 1-16), sometimes referred to as the *Oerlikon gun,* is located either forward or aft of the conning tower on the bridge deck. It is a rapid fire, recoil type of gun. In some instances the single mount has been replaced by twin mounts. Four 20-mm guns are carried by the fleet type submarine.

In addition to the armament described above, the fleet type submarine carries other small arms and pyrotechnics (Figure 1-17). Chief among these are two 30-caliber and four 50-caliber Browning machine guns, and one 45-caliber Thompson submachine gun.

2
DEFINITIONS AND PHRASEOLOGY

A. GENERAL DEFINITIONS

Introduction. The definitions contained in this chapter are exact meanings of the terms commonly used in reference to the modern submarine and its operation. These terms and explanations represent accepted interpretations and provide an understanding of the functions of the equipment.

2A1. Surface condition. A submarine is in *surface condition* when she has sufficient positive buoyancy to permit running on her main engines.

2A2. Diving trim. The term *diving trim* designates that condition of a submarine when it is so compensated that completing the flooding of the main ballast, safety, and bow buoyancy tanks will cause the vessel to submerge with neutral buoyancy and zero fore-and-aft trim.

2A3. Rigged for dive. A submarine is *rigged for dive* by so compensating the vessel and preparing the hull openings and machinery that the vessel can be quickly and safely submerged and controlled by flooding the main ballast tanks, using the diving planes, and operating on battery-powered main motors

2A4. Running dive. A *running dive* consists of submerging a submarine while running on battery power.

2A5. Stationary dive. A *stationary dive* consists of submerging a submarine without headway or sternway.

the fore-and aft and over-all weights have been so adjusted that the boat maintains the desired depth, on an even keel, at slow speed, with minimum use of the diving planes.

2A9. Compensation. *Compensation* is the process of transferring ballast, in the form of water, between the variable tanks, and between the variable tanks and sea, to effect the desired trim.

2A10. Main ballast tanks. Tanks that are provided primarily to furnish buoyancy when the vessel is in surface condition and that are habitually carried completely filled when the vessel is submerged, except tanks whose main volume is above the surface waterline, are known as *main ballast tanks.*

2A11. Variable ballast tanks. Ballast tanks that are not habitually carried completely filled when submerged and whose contents may be varied to provide weight compensation are known as *variable ballast tanks.* Variable ballast tanks are constructed to withstand full sea pressure.

2A12. Negative tank. The *negative tank* is a variable ballast tank providing negative buoyancy and initial down-angle. Submarines normally will operate submerged in neutral buoyancy and without trim when the negative tank is nearly empty. It is used to reduce the time required in submerging from surface condition, to reduce the time required to increase depth while operating submerged, and to prevent

2A6. Quick dive. A *quick dive* consists of rapidly submerging a submarine while running on main engines.

2A7. Submerged condition. This term designates a condition of a submarine in which all fixed portions of the vessel are completely submerged and the variable ballast is so adjusted that the submarine has approximately neutral buoyancy and zero fore-and-aft trim.

2A8. Final trim. *Final trim* is the running trim obtained after submerging, in which

broaching when decreasing depth. It may be blown or pumped.

2A14. Bow buoyancy tank. The *bow buoyancy tank* is a free-flooding, vent-controlled

tank with its main volume above the normal surface waterline. It is located in the extreme bow of the vessel and is formed of the plating of the superstructure. Its function is to provide reserve surface buoyancy, emergency positive buoyancy in the submerged condition, and to aid in surfacing.

2A15. Auxiliary tanks. The *auxiliary tanks* are variable ballast tanks located at or near the submerged center of buoyancy, and are used to vary the over-all trim of the boat.

2A16. Trim tanks. The *trim tanks* are the variable ballast tanks nearest the bow and stern of the boat and are used to provide fore-and-aft compensation.

2A17. Normal fuel oil tanks. Tanks designed solely for containing the engine fuel oil are known as *normal fuel oil tanks.*

2A18. Fuel ballast tanks. The *fuel ballast tanks* are designed to be utilized as fuel oil tanks for increased operating

below the breech of the torpedo tubes. The air and water from the poppet valves, incident to the firing of torpedoes, is discharged into this tank.

2A23. Fresh water tanks. The *fresh water tanks* contain potable water for drinking, cooking, and certain sanitary facilities.

2A24. Battery fresh water tanks. The *battery fresh water tanks* are storage tanks for the distilled water used in watering the main storage batteries.

2A25. Sanitary tanks. The *sanitary tanks* receive and store the ship's sanitary drainage until conditions permit overboard discharge.

2A26. WRT tanks. The WRT, or *water round torpedo*, tanks are variable ballast tanks, located in the forward and after torpedo rooms, for flooding or draining the torpedo tubes.

2A27. Main vents. The *main vents* are valves operated hydraulically, or by

range. When empty, they may be converted to main ballast tanks, providing additional freeboard and thereby increasing surface speed.

2A19. Expansion tank. The *expansion tank,* connected between the head box and the compensating water main, admits sea pressure to the fuel oil tanks. It receives any overflow from the fuel tanks resulting either from overfilling the fuel system or from temperature expansion. The bilges are pumped into this tank to prevent leaving an oil slick or polluting a harbor.

2A20. Collecting tank. The *collecting tank*, connected to the fuel oil tanks through the fuel transfer line, serves as a water and sediment trap for the fuel oil being transferred to the fuel pump.

2A21. Clean fuel oil tanks. The *clean fuel oil tanks* are storage tanks located within the pressure hull. They receive clean fuel oil from the purifiers and are the supply tanks from which the engines receive their clean fuel.

2A22. Poppet valve drain tank. The *poppet valve drain tank* is located under the platform deck of the torpedo room immediately

hand, for venting the main ballast tanks when flooding. They are located in the top of the risers of the main ballast tanks.

2A28. Emergency vents. The *emergency vents* are stop valves in the vent risers near the tank tops and are used in case of damage to the, main vents. They permit sealing the tank to prevent accidental flooding and also permit blowing the tank if desired.

2A30. Riding the vents. *Riding the vents* is a surface condition in which the main ballast tanks are prevented from completely flooding by the closed main vents which prevent the escape of air.

2A31. Flood Valves. *Flood valves* are hinged covers at the bottom of certain ballast tanks which may be opened to admit or expel sea water.

2A32. Flooding. Filling a tank through flood ports, open flood valves, or other sea connections, is known as flooding.

2A33. Blowing. *Blowing* a tank consists of expelling its contents by compressed air.

2A34. Pumping. *Pumping* a tank consists of using a pump to transfer liquid from the tank to sea, from sea to tank, or from one tank to another. The tanks must be vented during this operation.

2A35.Bow planes. The *bow planes* are horizontal rudders, or diving planes, extending from each side of the submarine near the bow.

2A36. Stern planes. The *stern planes* are horizontal rudders, or diving planes, extending from each side of the submarine near the stern.

B. STANDARD PHRASEOLOGY

2B1. General. Standard phraseology is the product of years of experience and has been developed to combine precision, brevity, and audibility. The following procedures have been approved for submarine communications, both airborne and over interior communication systems. Strict adherence to these procedures increases the speed of communications and reduces the chances of error and misunderstanding. The standard phrases, developed for the various activities of a submarine, are included in the chapter in which their use occurs.

bility and to minimize confusion. This is standard for the service, and should be followed invariably.

The numeral "0" is spoken as "Ze-ro" for all numerical data except ranges. In giving ranges, "0" is spoken as "Oh." When "00" occurs at the end of a number it is spoken as "Double-oh."

Examples: "Bearing too ze-ro ze-ro."
"Range fi-yiv oh double-oh."

a. Bearings and courses are spoken

NUMERAL	SPOKEN AS	NUMERAL	SPOKEN AS
0	Ze-ro or Oh (stress on both syllables of Zero)	5	FI-yiv (stress on first syllable)
1	Wun	6	Six
2	Too	7	Seven
3	Thuh-REE (stress on second syllable)	8	Ate
4	FO-wer (stress on first syllable)	9	Niner

2B2. Voice procedure. All messages should be spoken clearly and loudly enough to be heard above the noises and voices of the various compartments. Talk slowly and speak distinctly, do not run words together. Make the listener hear all you say the first time you say it.

Numerals. Exhaustive tests have demonstrated that numerals should be spoken in the following manner to provide intelligi-

as three separate digits.
Examples: "Bearing ze-ro zero thuh-ree."
"Steer course wun niner six."

b. Speed and torpedo depths are spoken as two separate digits.

Examples: "Speed ze-ro six and wun half knots."
"Set depth wun too feet."

c. Angle on the bow is spoken as a single compound number preceded by "port." or "starboard."

b. When an order has been executed, that fact is communicated to the originating station. *Example:* Statement

Example: "Angle on the bow port thirty fi-yiv."

d. Depth to keep, and *bubble*, or angle of the boat and angle on the planes, are spoken as separate digits.

Example: "Six fi-yiv feet, too degree up bubble, too zero degrees rise on the bow planes."

e. Time is spoken in standard Navy terminology .

Examples: "Ze-ro ze-ro thirty." "Ze-ro ate hundred." "Seventeen thirty fi-yiv." "Ze-ro niner ze-ro fi yiv."

Messages. a. Messages over a telephone or talk-back normally consist of two parts: 1) the *call* and 2) the *text*. There should be no pause between these parts for acknowledgment by the receiver.

Example: "After room, open outer doors aft."

b. When it is necessary to prevent misunderstanding, the station calling should identify itself immediately after the call.

Example: "Control, forward room: we heard a bumping noise along the hull !"

2B5. Acknowledgment. a. Each message should be acknowledged by an exact repetition. "Aye, aye" should not be used because it gives the originator no clue as to whether or not the message has been understood correctly.

Example: Message. "After room, open outer doors aft." Acknowledgment. "After room, open outer doors aft."

of execution. "Conning tower, the outer doors have been opened aft."

Acknowledgment. "Conning tower, the outer doors have been opened aft."

c. When a question cannot be answered immediately, it is acknowledged and the word "Wait" added. The question is answered as soon as the information is available.
Example: Message. "After Engine Room, how are the bilges ?"

Acknowledgment. "After Engine Room, how are the bilges? Wait."

Reply, after the information is obtained. "Control, six inches of water in the after engine room bilges." Acknowledgment. "Control, six inches of water in the after engine room bilges."

d. If the acknowledgment shows that the message has not been heard correctly, or if the originator himself decides to change the message, he says, "Belay that," and gives the correct form.

e. A repeat is requested whenever there is any doubt concerning the content of a message.

2B6. Emergency messages. In case of emergency, the station making announcement calls, "Silence on the line." All other stations cease talking until the emergency message has been completed.

2B7. Courtesy. The words "sir" and " please", and so forth, are not used on interior communication circuits. On a combat vessel, courtesy consists of making telephone messages as brief and efficient as possible.

C. COMMON ABBREVIATIONS

2C1. Acceptable abbreviations. In the box below are given some of the most frequently used abbreviations. They are time savers and should be used whenever possible. In the interest of uniformity throughout the Service they should be used exactly as they appear here.

ABBREVIATION	MEANING
W. S.	water pressure test for strength
W.T.	water pressure test for tightness
W.S.&T.	water pressure test for strength and tightness
A.S.	air pressure test for strength
A.T.	air pressure test for tightness
A.S.&T.	air pressure test for strength and tightness
O.S.	oil pressure test for strength
O.T.	oil pressure test for tightness
F.O.T.	fuel oil test for tightness
L.O.T.	lubricating oil test for tightness
O.S.&T.	oil pressure test for strength and tightness
MBT	main ballast tank
FBT	fuel ballast tank
NFOT	normal fuel oil tank
CFOT	clean fuel oil tank
NLOT	normal lubricating oil tank
WRT	water round torpedo
psi	pounds per square inch
hp	horsepower
rpm	revolutions per minute
shp	shaft horsepower

3
COMPARTMENTATION AND EXTERIOR INSTALLATIONS

A. COMPARTMENTATION

3A1. General. A modern submarine contains, in addition to the mechanisms required to operate it on the surface, a multitude of operating machinery and tanks required to enable it to dive, surface, and proceed submerged. This fact makes it one of the most compact vessels afloat. Yet the submarine is designed and arranged along simple and logical lines, and in spite of the seeming confusion of valves, lines, and apparatus, everything in the submarine is situated to insure the maximum of speed and efficiency. (See Figure A-1.)

The modern fleet type submarine consists of a superstructure and a hull surrounded for the most part by various fuel and water ballast tanks. The pressure hull, designed to withstand the sea pressure, houses most of the ship's machinery and provides the living quarters for the officers and the crew. It is divided into eight watertight compartments, separated by pressure bulkheads provided with watertight pressure resistant doors. The ninth compartment, the conning tower, in the shape of a cylinder placed on its side, is located above the control room and connects with the control room through the access hatch.

The compartments, in turn, are divided by means of the platform deck into upper and lower sections, which contain the spaces housing the various equipment and providing the necessary facilities for the submarine's officers and crew.

3A2. Forward torpedo room. The foremost compartment in the submarine spare torpedoes; torpedo loading hatch overhead; and torpedo handling equipment. Four of the torpedo tubes, Nos. 1, 2, 3 and 4, are above the platform deck, while tubes Nos. 5 and 6 are below the platform deck. The sonar gear, underwater log, and an access hatch to the escape trunk are also located in the forward torpedo room.

The following is a list of the more important equipment located in the forward torpedo room:

1. Hydraulic pump and ram for bow plane tilting
2. Hydraulic motor for windlass and capstan and bow plane rigging
3. Impulse charging manifolds (port and starboard)
4. Torpedo tube blow and vent manifold
5. Torpedo tube drain manifold
6. Torpedo gyro regulators
7. Blow and vent manifold for normal fuel oil tanks No. 1 and No. 2
8. Blow and vent manifold for No. 1 sanitary tank
9. Blow and vent manifold for fresh water tanks No. 1 and No. 2
10. Officers' head
11. Bunks and lockers for crew (usually at least 10 men)
12. High and low external compartment air salvage valves
13. Compartment air salvage (internal)
14. Oxygen flasks
15. Sound-powered telephones
16. Bow buoyancy vent operating gear
17. No. 1 main ballast tank vent

is the forward torpedo room (See Figure 3-1.), located between frames 16 and 35. The forward torpedo room contains six torpedo tubes in its forward bulkhead; torpedo racks on its port and starboard sides, immediately adjacent to the torpedo tubes for carrying

operating gear

22

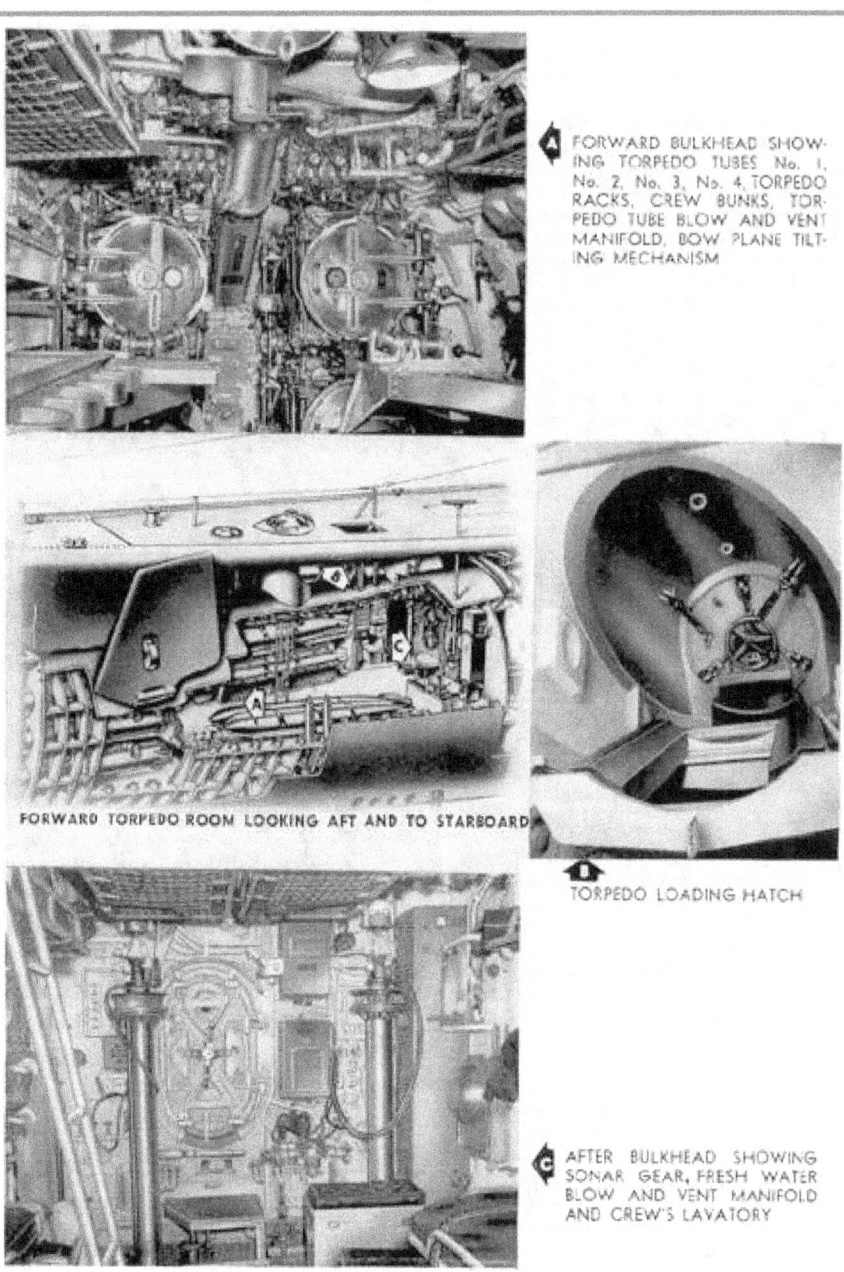

FORWARD BULKHEAD SHOW-ING TORPEDO TUBES No. 1, No. 2, No. 3, No. 4, TORPEDO RACKS, CREW BUNKS, TOR-PEDO TUBE BLOW AND VENT MANIFOLD, BOW PLANE TILT-ING MECHANISM

FORWARD TORPEDO ROOM LOOKING AFT AND TO STARBOARD

TORPEDO LOADING HATCH

AFTER BULKHEAD SHOWING SONAR GEAR, FRESH WATER BLOW AND VENT MANIFOLD AND CREW'S LAVATORY

Figure 3-1. Forward torpedo room.

STARBOARD SIDE OF AFTER BULK-
HEAD AT FRAME 47 SHOWING
DOOR TO CONTROL ROOM, EX-
TERNAL COMPARTMENT SAL-
VAGE CONNECTION VALVE,
CELL TESTING PANEL, EMER-
GENCY LIGHT, AND BULKHEAD
FLAPPER

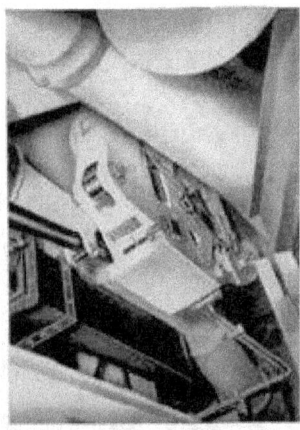

FORWARD BATTERY COMPARTMENT—
LOOKING AFT TO STARBOARD

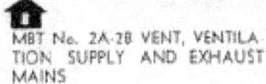

MBT No. 2A-2B VENT, VENTILA-
TION SUPPLY AND EXHAUST
MAINS

GENERAL VIEW OF BATTERY
SPACE SHOWING THE EXHAUST
DUCTS AND TOP OF BATTERY
CELLS

Figure 3-2. Forward battery compartment.

3A3. Forward battery compartment.
(See Figure 3-2.) The forward torpedo
room is separated from the forward
battery compartment by a watertight
bulkhead and door. The lower part of
the forward battery compartment

6. Auxiliary power switchboard

7. Main ballast tanks Nos. 2A and 2C

Port side forward-aft:

houses the 126 forward battery cells. The upper section of the compartment contains the officers' quarters, the chief petty officers' quarters, and the yeoman's office. The entire forward battery compartment is located between frames 35 and 47. The officers' quarters provide wardroom and staterooms for the ship's officers, sleeping accommodations for the chief petty officers, and officers' showers and pantry.

Also located in the forward battery compartment are:

1. Ventilation supply lines
2. Ventilation exhaust lines
3. Bulkhead flappers
4. Main ballast tanks No. 2A-2B vent operating gear
5. Battery blowers
6. External salvage air connections (one high and one low)
7. Compartment internal salvage air connections
8. Oxygen flask
9. Sound-powered telephones

3A4. Control room. (See Figures 3-3, 3-4, and 3-5.) Going aft, the compartment immediately adjoining the forward battery compartment is called the *control room*. It is located between frames 47 and 58 and is the main control center of the submarine. The control room contains:

Starboard side forward-aft:

1. Electric circuits and switchboard (IC switchboard)
2. High-pressure air manifold
3. 225-pound air manifold

1. Gun access batch
2. Oil supply tank
3. Signal ejector
4. Hydraulic air-loading manifold
5. Negative tank inboard vent
6. Hydraulic manifold
7. Bow and stern plane diving station
8. Trim manifold
9. Main ballast tanks Nos. 2B and 2D emergency vents

Centerline forward-aft:

1. Steering stand
2. Master gyro
3. Pump room hatch
4. Periscope wells
5. Sound-powered telephones
6. Radar mast

In the overhead:

1. High and low external compartment air salvage
2. Internal compartment air salvage
3. Ventilation supply and exhaust lines
4. Bulkhead flappers
5. Oxygen flasks

In the pump room:

1. Hand and hydraulic operating gear for negative flood
2. High pressure air compressors
3. Low-pressure blower
4. Drain pump
5. Trim pump
6. Vacuum pump
7. IMO pumps
8. Hydraulic accumulator

4. 600-pound main ballast tanks blowing manifold
5. 10-pound main ballast tanks blowing manifold

9. Air-conditioning machines
10. Refrigeration machine
11. IC motor generators

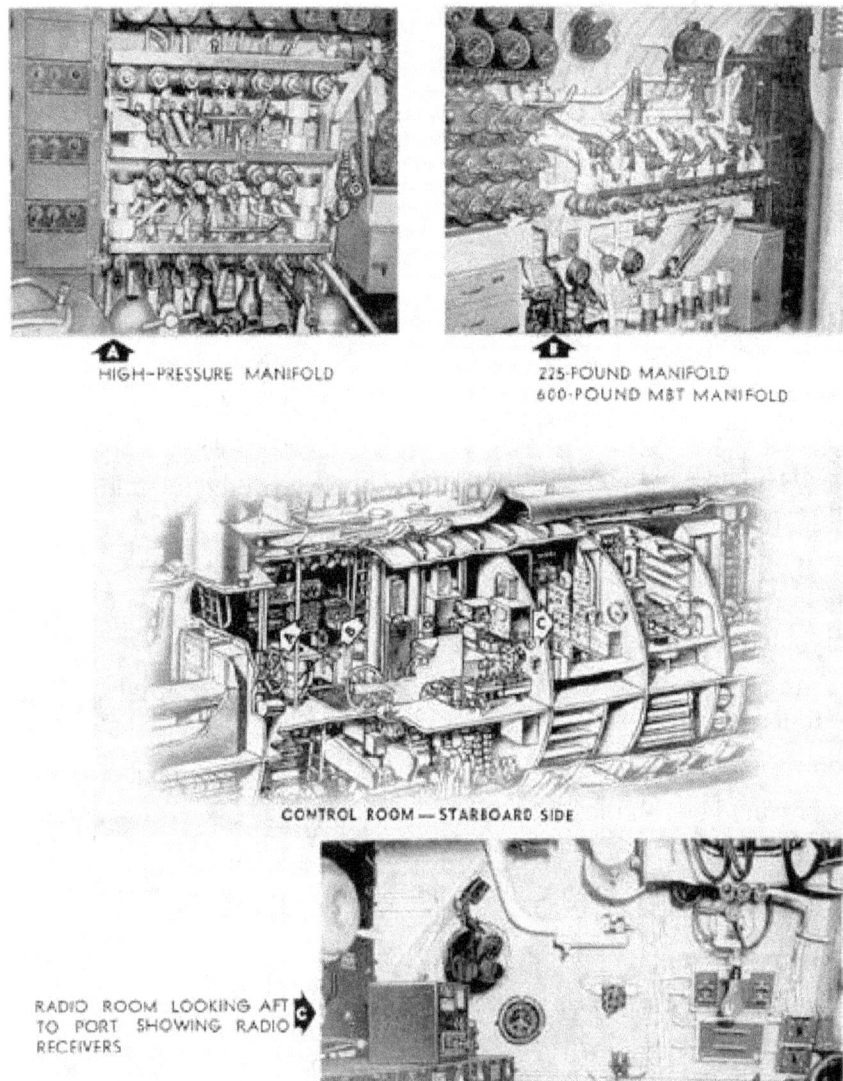

HIGH-PRESSURE MANIFOLD

225-POUND MANIFOLD
600-POUND MBT MANIFOLD

CONTROL ROOM—STARBOARD SIDE

RADIO ROOM LOOKING AFT TO PORT SHOWING RADIO RECEIVERS

Figure 3-3. Control room (starboard side) and radio room

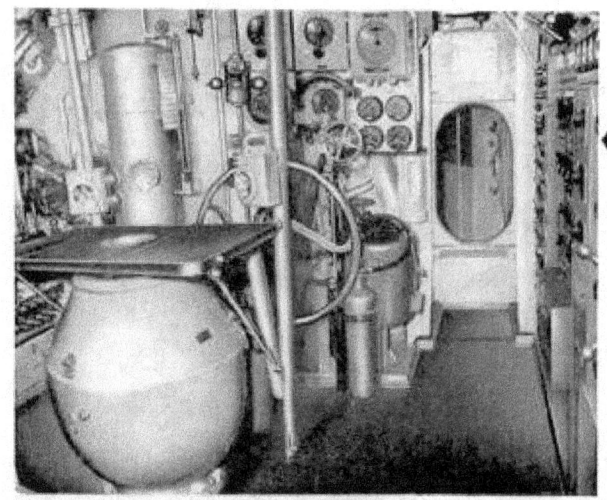

FORWARD BULKHEAD SHOWING GYRO COMPASS, CONTROL ROOM STEERING STAND, AUXILIARY GYRO COMPASS, HYDRAULIC MANIFOLD, HYDRAULIC OIL TANK, CHART TABLE, AND DOOR TO OFFICERS' QUARTERS

CONTROL ROOM — LOOKING AFT

GENERAL VIEW — AFTER STARBOARD SIDE OF CONTROL ROOM — SHOWING AIR MANIFOLDS, AIR PRESSURE GAGES, AND AFTER BULKHEAD WITH DOOR LEADING TO CREW'S QUARTERS

PORTSIDE AFTER CORNER OF CONTROL ROOM SHOWING TRIM MANIFOLD, FATHOMETER INDICATOR, TRIM TANK GAGES AND TRIM PUMP CONTROL

Figure 3-4. Control Room

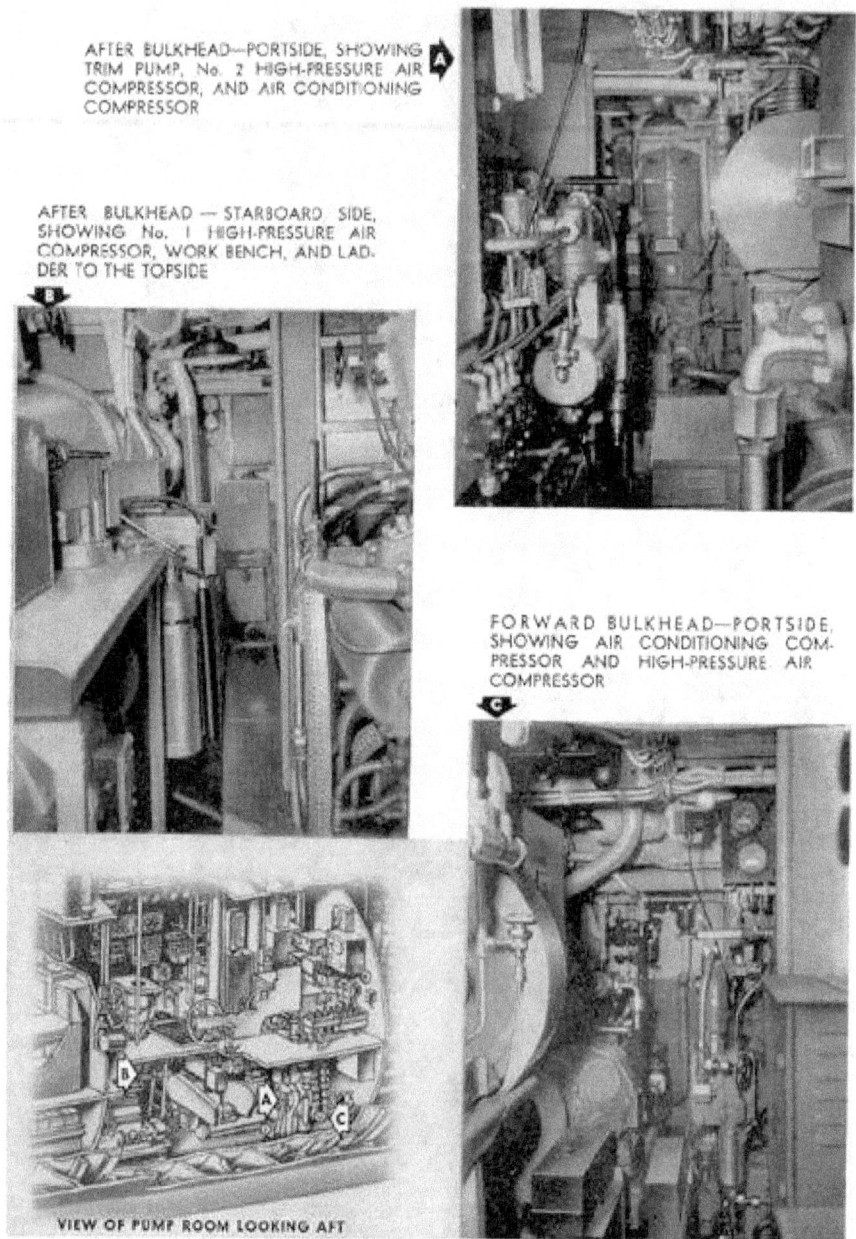

AFTER BULKHEAD—PORTSIDE, SHOWING TRIM PUMP, No. 2 HIGH-PRESSURE AIR COMPRESSOR, AND AIR CONDITIONING COMPRESSOR

AFTER BULKHEAD — STARBOARD SIDE, SHOWING No. 1 HIGH-PRESSURE AIR COMPRESSOR, WORK BENCH, AND LADDER TO THE TOPSIDE

FORWARD BULKHEAD—PORTSIDE, SHOWING AIR CONDITIONING COMPRESSOR AND HIGH-PRESSURE AIR COMPRESSOR

VIEW OF PUMP ROOM LOOKING AFT

Figure 3-5. Pump Room

The after section of the control room is occupied by the radio room which houses the transmitting and receiving radio apparatus, and radio direction finder. The engine induction and ship's supply outboard valve operating gear is located in the radio room overhead.

rooms and the ammunition magazine. Above the platform deck, the after battery compartment contains the crew's galley, mess hall, and the crew's sleeping quarters.

The more important equipment in the after battery compartment consists of

Access to the radio room is from the control room, which is in turn is accessible from the officers' quarters passageway forward, crew's mess hall aft, and the conning tower access hatch overhead.

3A5. Conning tower. The compartment immediately above the control room is the *conning tower.* (See Figure 3-6.) It is the main navigation and firing control station for the submarine. The conning tower contains the periscopes and the periscope hoist equipment, the radio direction finder, the sonar equipment, the radar equipment, the torpedo data computer (TDC), the gyro repeater, the conning tower steering stand, and the various pressure gages and indicators.

The conning tower connects with the control room through a watertight hatch. This is designated as the *lower conning tower hatch.* The upper conning tower hatch provides access to the bridge from the conning tower. The conning tower also has a ventilation exhaust connection and its own air-conditioning coil. Since the conning tower is the commanding officer's battle station, all communication lines include the conning tower in their circuits.

3A6. After battery compartment. The compartment aft of the control room is the after battery compartment (See Figure 3-7.), located between frame 58 and frame 77. It houses the after battery with its 126 cells below the platform deck. The battery cells are connected to the exhaust system by ducts leading to the battery blowers. The forward end of the after battery space, below decks,

emergency vents for safety tank, FBT No. 3 and No. 4; hand-operated flood valves for FBT No. 3 and No. 4; main vent operating gear for safety tanks, MBT No. 2C and 2D, FBT No. 3A and 3B, FBT No. 4A and 4B, and FBT No. 5A and 5B; the hydrogen-detecting apparatus; the blow and vent manifold for fuel ballast tanks No. 3A and 3B, 4A and 4B. The galley equipment, scuttlebutt, battery exhaust blowers, and the radio receiver for the crew are also located in the after battery compartment.

The after part of the after battery compartment contains the crew's bunks (usually 36) with individual lockers for each bunk and a medicine locker. Separated from it by a non-watertight bulkhead and a door are the crew's head, showers, washing machine, and lavatories.

The after battery compartment is provided with an external high and low compartment salvage air valve.

3A7. Forward engine room. The forward engine room (See Figure 3-8.) is located between frames 77 and 88, and houses No. 1 and No. 2 main engines. The main engines extend from below the platform deck into the engine room space, with No. 1 main engine on the starboard side and No. 2 main engine on the port side. The main generators, No. 1 and No. 2, are below decks aft of the main engines beneath the platform deck, and are directly connected to them.

The forward engine room houses the vapor compression distillers, fuel oil pump (standby), lubricating oil pump (standby), engine air inboard induction hull valve, and various main engine starting and stopping controls.

also contains the cool and the
refrigerating

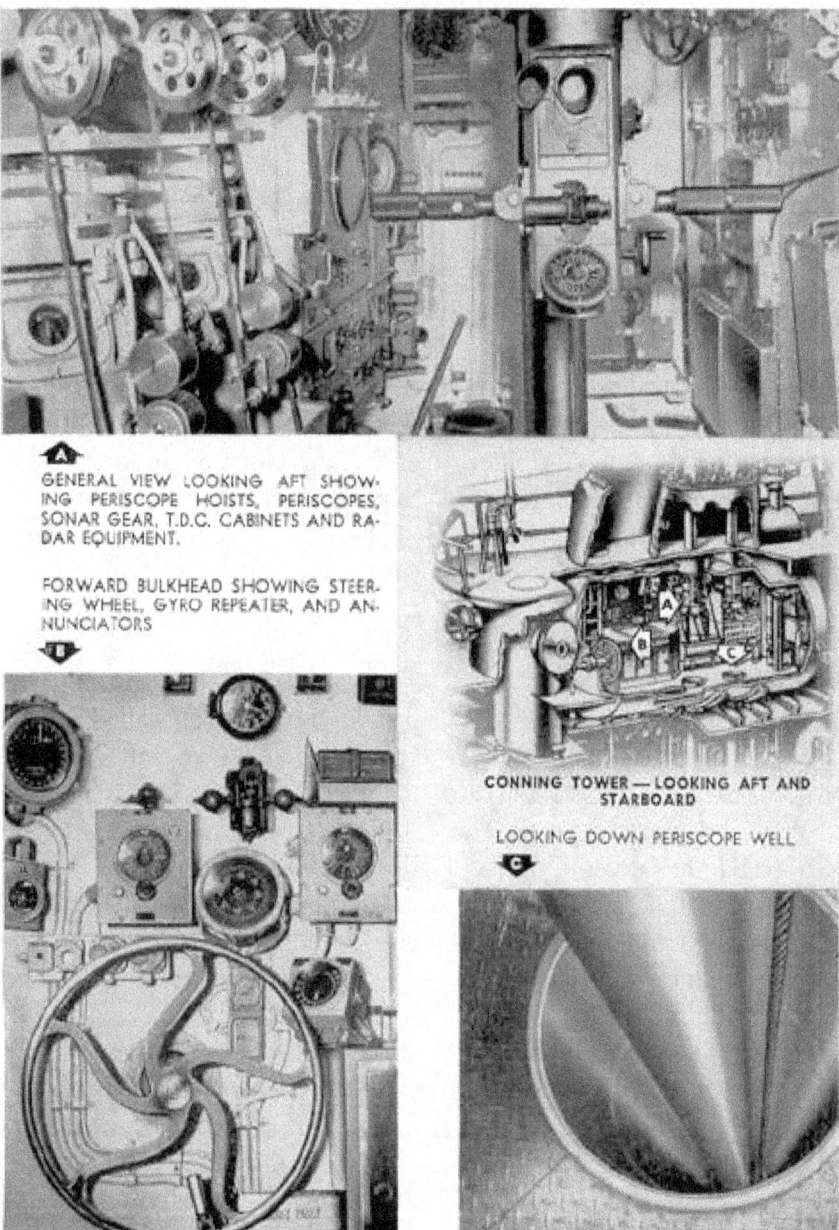

Figure 3-6. Conning tower

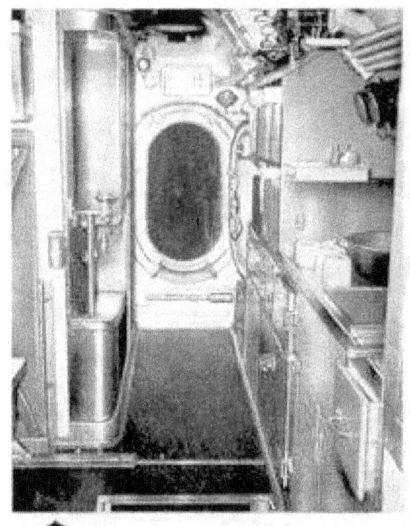

🔺 CREW'S MESS HALL—LOOKING FOR WARD—SHOWING FORWARD BULKHEAD WITH DOOR TO CONTROL ROOM, SCUTTLEBUTT, HATCH TO STORE ROOMS

🔺 CREW'S MESS HALL—LOOKING AFT—SHOWING DOOR TO CREW'S SLEEPING QUARTERS, MAIN DECK ACCESS HATCH, BENCHES AND TABLES

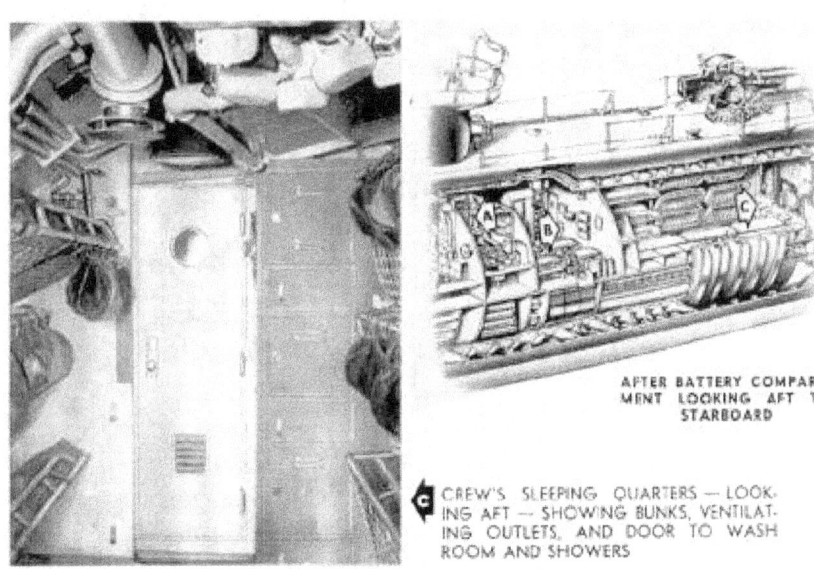

AFTER BATTERY COMPARTMENT LOOKING AFT TO STARBOARD

🔺 CREW'S SLEEPING QUARTERS—LOOKING AFT—SHOWING BUNKS, VENTILATING OUTLETS, AND DOOR TO WASH ROOM AND SHOWERS

Figure 3-7. After battery compartment and crew's quarters.

31

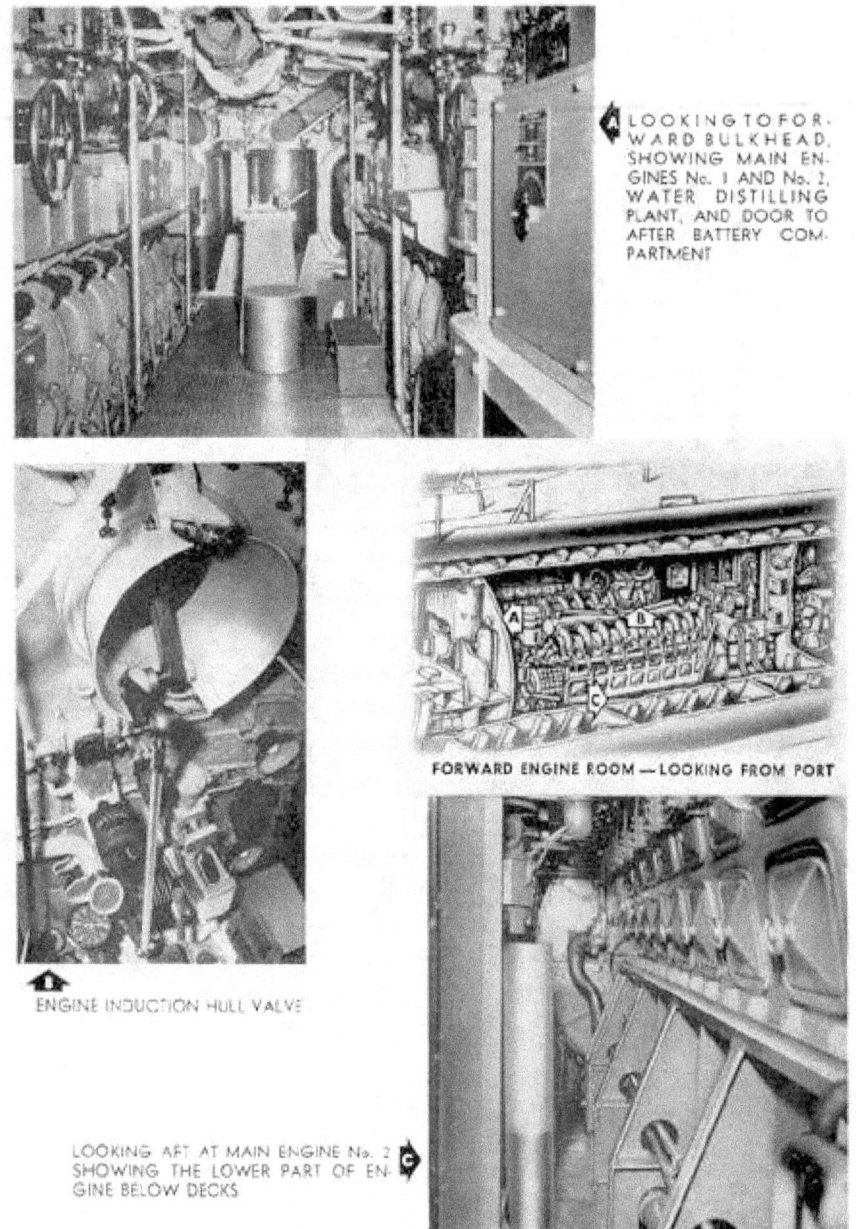

A LOOKING TO FORWARD BULKHEAD, SHOWING MAIN ENGINES No. 1 AND No. 2, WATER DISTILLING PLANT, AND DOOR TO AFTER BATTERY COMPARTMENT

FORWARD ENGINE ROOM — LOOKING FROM PORT

B ENGINE INDUCTION HULL VALVE

LOOKING AFT AT MAIN ENGINE No. 2 SHOWING THE LOWER PART OF ENGINE BELOW DECKS C

Figure 3-8. Forward engine room

32

The forward engine room also contains the following:

1. High and low external compartment air salvage connection

2. Internal compartment air salvage connection

3. Hull ventilation supply valve

and 3-11.) The upper section of the maneuvering room, frames 99 to 107), contains the maneuvering control stand, indicators, gages, lathe, crew's head, auxiliary switchboard, remote control for engine shutdown, oxygen flask, high and low external compartment air salvage, internal compartment air salvage, maneuvering room induction hull valve,

4. Supply blower

5. Exhaust blower

6. Bulkhead flappers

7. Oxygen flask

8. Sound-powered telephone

9. Fuel oil purifier

10. Lubricating oil purifier

3A8. After engine room. The after engine room (See Figure 3-9), located between frames 88 and 99, is similar in many respects to the forward engine room. It houses the No. 3 and No. 4 main engines and main generators.

In addition, the after engine room houses the auxiliary generator and the auxiliary diesel engine, both of which are located entirely below the platform deck.

The upper space of the after engine room houses the after engine air inboard induction hull valve, the high and low external compartment air salvage connection, the internal compartment air salvage valve, the lubricating oil and the fuel oil purifiers and the pumps. The compartment also contains:

1. Standby fuel oil pump

2. Standby lubricating oil pump

3. Air-conditioning unit

4. Hull supply lines (ventilation)

5. Access hatch

6. Oxygen flask

7. Sound-powered telephone

8. Bulkhead flappers

3A9. Maneuvering room. (See Figures 3-10

hull supply lines, and bulkhead flappers.

The lower part of the maneuvering room is called the motor room and has the Nos. 1, 2, 3, and 4 main motors. The four main motors are directly connected with the two reduction gears. The main motors Nos. 1 and 3 are directly connected with the reduction gear No. 1, and the main motors Nos. 2 and 4 are directly connected with reduction gear No. 2. The reduction gears in turn are connected with the propeller shafts.

The circulating water pumps and the lubricating oil pumps are also located in the motor room.

3A10. After torpedo room. The aftermost compartment on the submarine is the after torpedo room (See Figure 3-12), located between frame 107 and 125. Unlike the forward torpedo room, it contains only four torpedo tubes in its after bulkhead. However, it also has the torpedo racks, torpedo handling equipment, and spare torpedoes.

The after torpedo room has one impulse charging manifold, torpedo tube blow and vent manifold, torpedo gyro regulators, and the torpedo tube firing indicator and controls. It also contains the escape and rescue hatch, the torpedo loading hatch overhead, the hydraulic steering rams and pump, the stern place tilting mechanism, the torpedo tube drain manifold, the crew's bunks and lockers, the ventilation supply line, the bulkhead flappers, the oxygen flasks, the emergency air connection for escape hatch, the high and low external air salvage connections, the internal compartment air salvage, and a sound-powered telephone.

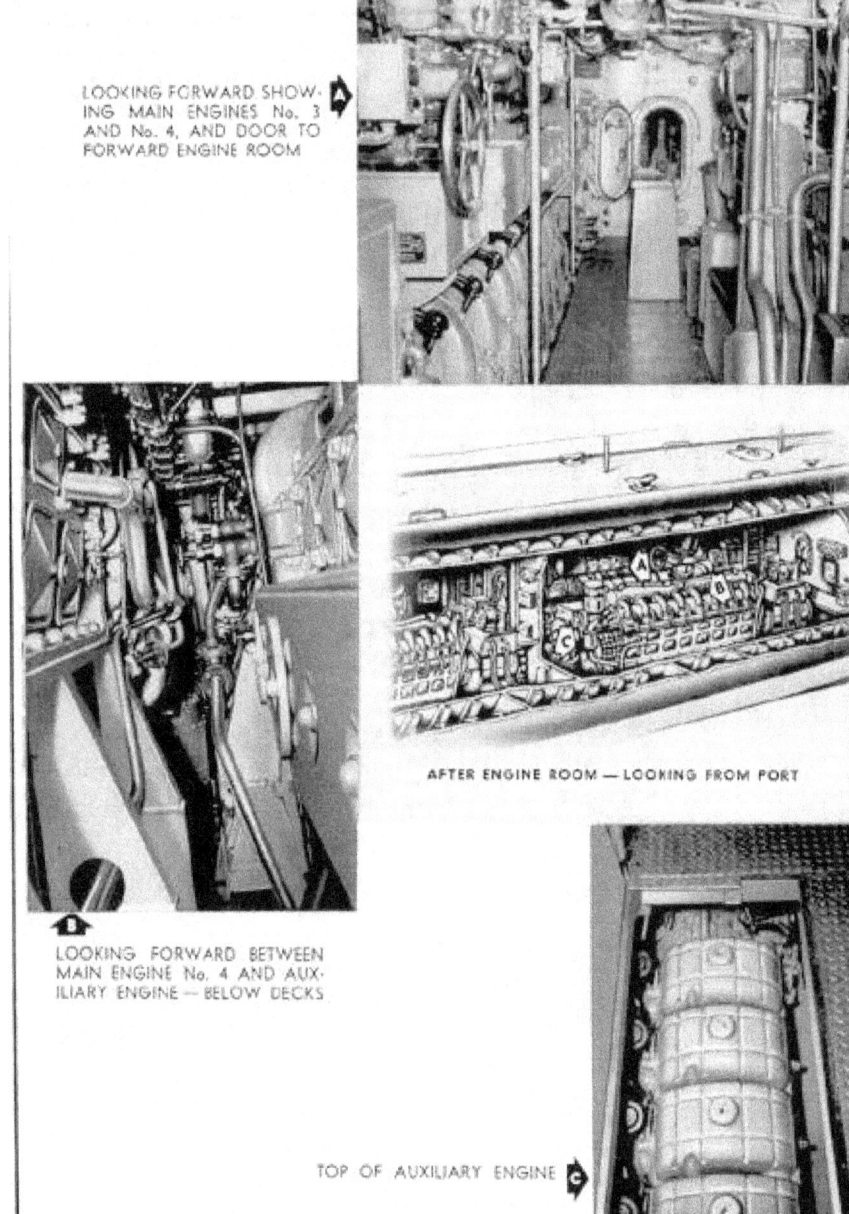

Figure 3-9. After engine room

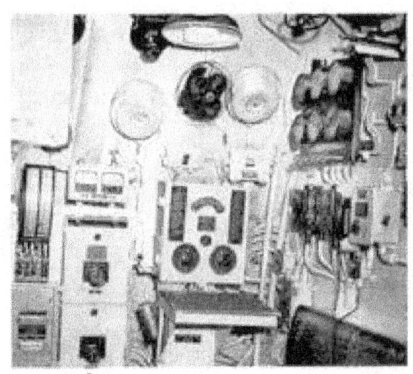

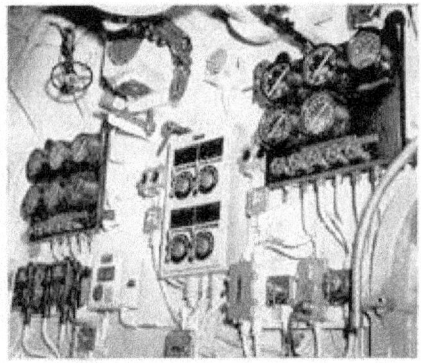

A STARBOARD AFTER CORNER SHOWING AIRFLOW INDICATOR, HYDROGEN DETECTOR

B AFTER BULKHEAD SHOWING MAIN MOTOR AND REDUCTION GEAR TEMPERATURE AND PRESSURE GAGES, ENGINE ORDER TRANSMITTER

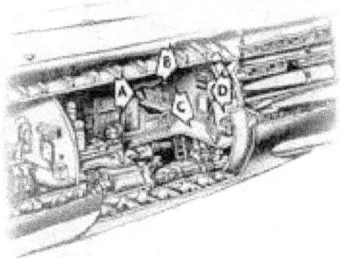

MANEUVERING ROOM—LOOKING AFT AND TO STARBOARD

C MAIN PROPULSION CONTROL LEVERS

D MAIN PROPULSION CONTROL

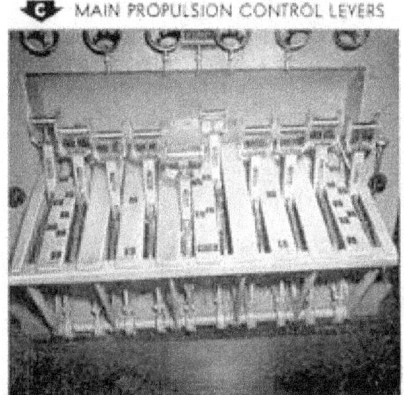

Figure 3-10. Maneuvering room above platform deck

35

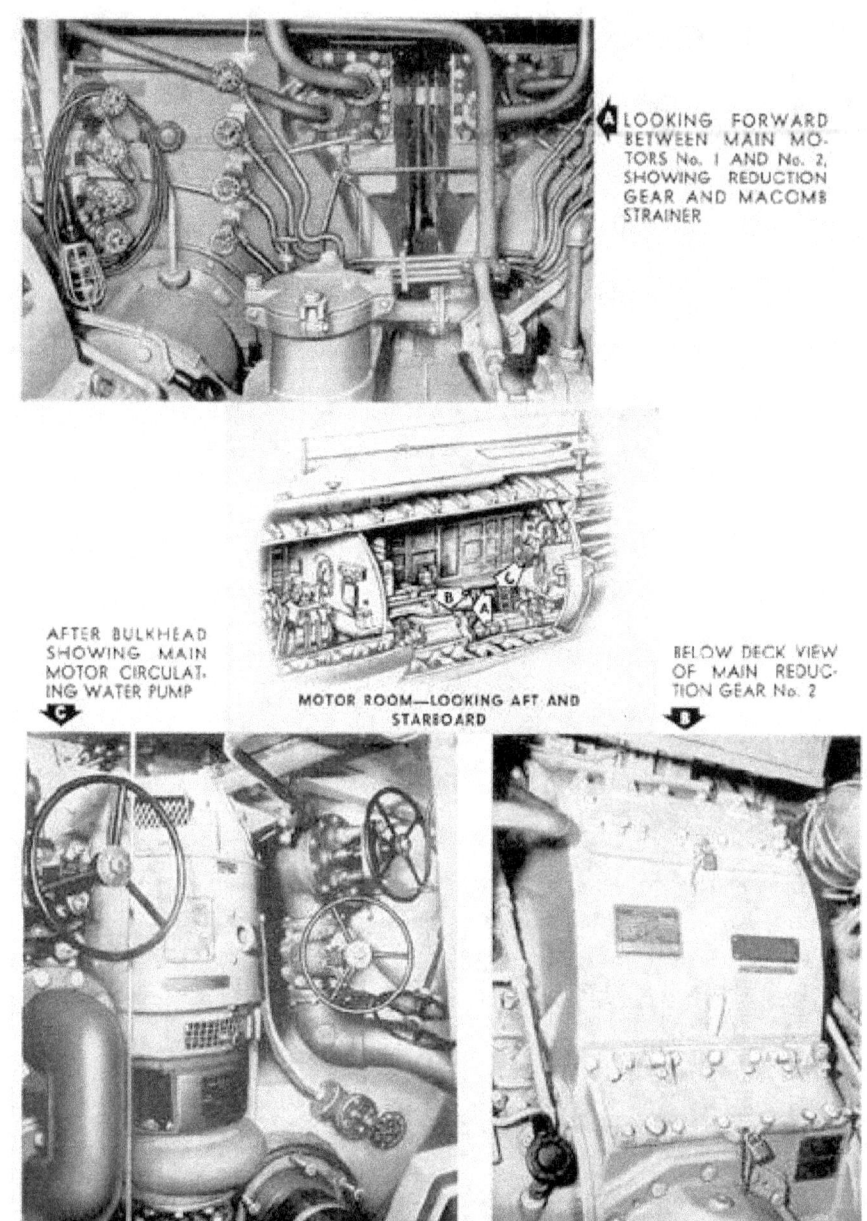

A LOOKING FORWARD BETWEEN MAIN MOTORS No. 1 AND No. 2, SHOWING REDUCTION GEAR AND MACOMB STRAINER

AFTER BULKHEAD SHOWING MAIN MOTOR CIRCULATING WATER PUMP C

MOTOR ROOM—LOOKING AFT AND STARBOARD

BELOW DECK VIEW OF MAIN REDUCTION GEAR No. 2 B

Figure 3-11. Motor room

36

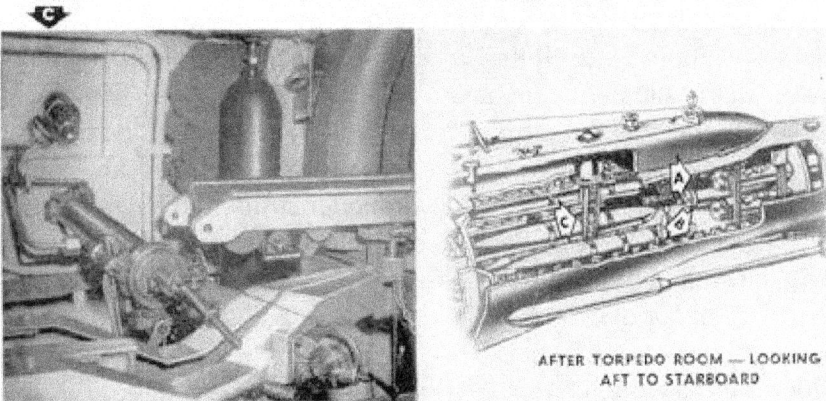

AFTER BULKHEAD SHOWING TORPEDO TUBES No. 7 AND No. 8, TORPEDO TUBE BLOW AND VENT MANIFOLD, GYRO REGULATOR

PORTSIDE AFT — SHOWING PORT STEERING RAM, MAIN RAM CUT-OUT MANIFOLD, TORPEDO TUBE INDICATOR LIGHTS

PORTSIDE AFT — SHOWING SIGNAL EJECTOR AND TOPEDO RACK

AFTER TORPEDO ROOM — LOOKING AFT TO STARBOARD

Figure 3-12. Maneuvering room above platform deck

37

B. EXTERIOR INSTALLATIONS

3B1. General. The exterior view of the submarine presents a very low silhouette. This is due primarily to the fact that the submarine is designed to have a low center of gravity for gun and after ammunition ready locker, while the forward section of the bridge may have either a 20-mm or a 40-mm antiaircraft gun, depending upon the particular ship. Two ammunition ready

stability and is normally two-thirds submerged as she rides on the surface.

The exterior hull of the submarine has a cylindrical shape, which gradually tapers forward of frame 35 and aft of frame 107, becoming the bow of the superstructure and the rounded stern. (See Figure A-2.)

The superstructure deck, called the *main deck,* extends virtually from the tip of the bow to frame 124 near the stern. The deck is generally level. Beginning about the midship section it rises gradually in the direction of the bow, to a height of approximately 12 feet above the water line. The freeboard of the after end of the main deck is about 4 feet.

The main deck is attached to the exterior hull by means of the framing and rounded sides. Limber holes in the sides allow sea water to enter all the hollow spaces in the superstructure and the deck when diving, and drain off when the submarine is surfacing.

The midship section of the main deck is occupied by the conning tower, which is surmounted by the bridge deck, with periscope shears, periscopes, radio compass loop, and radar antenna.

The after section of the bridge deck contains the ship's pelorus, one 20-mm anti-aircraft

lockers are located in the lower part of the conning tower superstructure, one forward and one aft. The gun access trunk is located forward and one aft. The gun access trunk is located forward of the conning tower and is provided with a hatch opening onto the main deck.

The forward section of the deck contains the 5"/25 wet type gun, galley access hatch, after engine room access hatch, after torpedo loading hatch, after rescue and escape hatch, marker buoy, and capstan.

The bow is equipped with six torpedo tube shutters, three on the port and three on the starboard side, and the bow diving planes.

The underside of the hull contains ballast tank flooding ports and underwater sound heads.

The after end of the ship, on the underside, is equipped with the four stern torpedo tubes, two on the port and two on the starboard side, port and starboard propeller struts, propellers, stern diving planes, and the rudder.

4
TANK ARRANGEMENTS

A. TANKS

4A1. General. In a submarine, the principal ballast is water. Therefore, the arrangement of tanks built into the ship establishes the points at which water ballast may be concentrated. It is the arrangement of these tanks that makes possible controlled diving and surfacing and the maintenance of diving trim at any depth. The arrangement of the tanks, with respect to the center of buoyancy, establishes the lever arm for maintaining fore and aft balance and athwartship stability. Figure A-3 shows, in schematic form, the general arrangement of the tanks within a submarine.

The water ballast tanks are divided into four main groups: the main ballast tanks, the variable ballast tanks, the special ballast tanks, and the fuel oil ballast tanks.

4A2. Main ballast tanks. The main ballast tanks group consists of four groups, which are further subdivided into ten tanks, as follows:

TANK	CAPACITY
1. MBT No. 1	49.17 tons sea water
2. MBT Nos. 2A, 2B, 2C and 2D	129.03 tons sea water (4 tanks)
3. MBT Nos. 6A, 6B, 6C and 6D	141.60 tons sea water (4 tanks)
4. MBT No 7	39.09 tons sea water

4A3. Variable ballast tanks. The second group of water tanks is the variable ballast tank group which is composed of six tanks as follows:

TANK	CAPACITY
1. Forward trim tank	24.31 tons sea water
2. Forward WRT tank	4.94 tons sea water
3. Auxiliary ballast tank No. 1	30.77 tons sea water
4. Auxiliary ballast tank No. 2	30.77 tons sea water
5. After trim tank	19.97 tons sea water
6. After WRT tank	5.06 tons sea water

4A4. Special ballast tanks. The safety, negative, and bow buoyancy tanks are classified as special ballast tanks. Each of these tanks has special blowing arrangements and a special purpose, which is described in detail in later sections of this chapter.

TANK	CAPACITY
1. Safety tank	23.23 tons sea water
2. Negative tank	7.51 tons sea water
3. Bow buoyancy tank	31.69 tons sea water

4A5. Fuel Ballast tanks. There are three fuel ballast tanks divided into A and B sections which are connected together through limber holes in the vertical keel plating. The tanks are as follows:

TANK	CAPACITY
1. Fuel ballast tanks Nos. 3A and 3B	19,196 gallons
2. Fuel ballast tanks Nos. 4A and 4B	24,089 gallons
3. Fuel ballast tanks Nos. 5A and 5B	19,458 gallons

The fuel ballast tanks normally carry fuel oil. When not being used as fuel ballast tanks, they may be used as main ballast tanks.

4A6. Additional tanks. In addition to the above-named water ballast tanks, there are the normal fuel oil tanks, collecting tank, expansion tank, clean fuel oil tank, normal lubricating oil tank, reserve lubricating oil tank, main sump tanks, reduction gear sump tanks, fresh water tanks, emergency fresh water tanks, battery fresh water tanks and sanitary tanks. The capacity of these tanks is given in the following table.

TANK	CAPACITY
Normal fuel oil tank group:	
1. NFOT No. 1	11,401 gallons
2. NFOT No. 2	13,122 gallons
3. NFOT No. 6	15,201 gallons
4. NFOT No. 7	10,054 gallons
5. Collecting tank	2,993 gallons
6. Expansion tank	2,993 gallons
Clean fuel oil tank group:	
1. CFOT No. 1	611 gallons
2. CFOT No. 2	618 gallons
Normal lubricating oil tank group:	
1. NLOT No. 1	1,475 gallons
2. NLOT No. 2	924 gallons
3. NLOT No. 3	1,073 gallons
4. Reserve lube oil tank	1,201 gallons

TANK	CAPACITY
Main engine sump tank group:	

1. Main engine sump No. 1	382 gallons
2. Main engine sump No. 2	382 gallons
3. Main engine sump No. 3	382 gallons
4. Main engine sump No. 4	382 gallons
5. Reduction gear sump No. 1	165 gallons
6. Reduction gear sump No. 2	165 gallons
Fresh water tank group:	
1. Fresh water tank No. 1	980 gallons
2. Fresh water tank No. 2	980 gallons
3. Fresh water tank No. 3	973 gallons
4. Fresh water tank No. 4	973 gallons
5. Emergency fresh water tanks	276 gallons (total)
a. 2 tanks forward torpedo room	276 gallons (total)
b. 1 tank control room	18 gallons
c. 1 tank maneuvering room	8 gallons
d. 1 tank aft torpedo room	180 gallons
Battery water tanks:	1,208 gallons (total)
1. Battery water tanks Nos. 1 and 2	152 gallons (each)
2. Battery water tank No. 3	143 gallons
3. Battery water tank No. 4	157 gallons
4. Battery water tank Nos. 5 and 6	152 gallons (each)
5. Battery water tank No. 7	157 gallons
6. Battery water tank No. 8	143 gallons
Sanitary tanks:	
1. Sanitary tank No. 1	1.66 tons or 434 gallons
2. Sanitary tank No. 2	2.57 tons or 673 gallons

41

4A7. Test pressure and data. The tank and groupings, together with their capacities outlined in Sections 4A2 and 4A6 inclusive, are those tanks which are designed to contain liquids under any normal or emergency condition of operation of the vessel. These tanks are subjected to the individual tests listed in the following chart:

TANK	TYPE OF TEST
1. MBT No. 1	A. S. & T. 15 psi Tests made
2. MBT Nos. 2A and 2B, 2C and 2D	A. S. & T. 15 psi before flood

TANK	TYPE OF TEST
3. MBT Nos. 6A and 6B, 6C and 6D	A. S. & T. 15 psi ports are cut
4. MBT No. 7	A. S. & T. 15 psi into tank.
5. Forward trim tank	W. S. & T. Test depth
6. Forward WTB tank	W. S. & T. Test depth
7. Auxiliary tank No. 1	W. S. & T. Test depth
8. Auxiliary tank No. 2	W. S. & T. Test depth
9. After trim tank	W. S. & T. Test depth
10. Safety tank	W. S. & T. Test depth
11. Negative tank	W. S. & T. Test depth
12. Bow buoyancy	W. S. & T.
13. FBT Nos. 3A and 3B	W. S. & T. 102 ft. head to keel
14. FBT Nos. 4A and 4B	W. S. & T. 102 ft. head to keel
15. FBT Nos. 5A and 5B	W. S. & T. 102 ft. head to keel
16. NFOT No. 1	W. S. & T. 102 ft. head to keel
17. NFOT No. 2	W. S. & T. 102 ft. head to keel
18. NFOT No. 6	W. S. & T. 102 ft. head to keel
19. NFOT No. 7	W. S. & T. 102 ft. head to keel
20. Collecting tank	W. S. & T. 102 ft. head to keel
21. Expansion tank	W. S. & T. 102 ft. head to keel
22. CFOT No. 1	W. S. & T. 60 ft. head to keel
23. CFOT No. 2	W. S. & T. 60 ft. head to keel
24. NLOT No. 1	W. S. & T. 35 ft. head to keel
25. NLOT No. 2	W. S. & T. 35 ft. head to keel
26. NLOT No. 3	W. S. & T. 35 ft. test depth
27. Reserve lube oil tank	W. S. & T. 35 ft. head to keel
28. Main engine sump No. 1	W. S. & T. 35 ft. head to keel
29. Main engine sump No. 2	W. S. & T. 35 ft. head to keel

TANK	TYPE OF TEST
30. Main engine sump No. 3	W. S. & T. 35 ft. head to keel
31. Main engine sump No. 4	W. S. & T. 35 ft. head to keel
32. Reduction gear sump No. 1	W. S. & T. test depth
33. Reduction gear sump No. 2	W. S. & T. test depth

34. Fresh water tank No. 1	A. S. & T. 18 psi
35. Fresh water tank No. 2	A. S. & T. 18 psi
36. Fresh water tank No. 3	A. S. & T. 18 psi
37. Fresh water tank No. 4	A. S. & T. 18 psi
38. Emergency fresh water tank No.	A. S. & T. 10 psi
39. Battery water tanks Nos. 1 and 2	A. S. & T. 18 psi
40. Battery water tank No. 3	A. S. & T. 18 psi
41. Battery water tank No. 4	A. S. & T. 18 psi
42. Battery water tanks Nos. 5 and 6	A. S. & T. 18 psi
43. Battery water tank No. 7	A. S. & T. 18 psi
44. Battery water tank No. 8	A. S. & T. 18 psi
45. Sanitary tank No. 1	W. S. & T. test depth
46. Sanitary tank No. 2	W. S. & T. test depth

B. WATER BALLAST TANKS

4B1. Purpose of water ballast tanks. The water ballast tanks include the main ballast tanks, the variable ballast tanks, and the special ballast tanks. The purpose of these tanks can best be defined by illustration. Assume that a new 1,500-ton submarine is making its initial dive, and that this trim dive is to be a stationary dive.

The ship has a surface displacement of 1,500 tons and draws 14 feet of water. When fuel oil and lube oil tanks are completely filled, she draws 15 feet 6 inches of water and is ready for her trim dive. The ship is on the surface and weighs 1,750 tons; this is the designed weight plus oil, stores, and crew. The submarine is ready for sea. The problem is to take on weight enough so that the ship will submerge to a depth at which the waterline will be even with the periscope shears. (With a draft of 15 feet 6 inches, the waterline is 31 feet 6 inches from the periscope shears.)

The weight taken on is water, and it is flooded into tanks. The air, of course, is vented off the tanks as the water flows in. First, the large tanks, known as *main ballast tanks,* are flooded. These tanks hold 359 tons of sea water. (See Section 4A2.) The submarine now displaces 2,109 tons and draws approximately 22 feet of water. The main deck is not awash, since there are approximately 2 feet from waterline to deck. The ship still has plenty of positive buoyancy. Since the bow buoyancy tank vent has been open during this operation, allowing this free-flooding tank to take on ballast as the ship submerges, it is necessary to add to the displacement the weight of water taken on by the bow buoyancy tank (which belongs to the special ballast tank group).

43

This gives a new total displacement of 2,141 tons (2,109 tons plus 32 tons).

Simultaneously with the flooding of bow buoyancy, the safety tank also in the special ballast tank group, is flooded. This tank holds 23 tons of water, giving a total displacement of 2,173 tons and a draft of 24 feet. The decks are just awash, and some positive buoyancy is still retained, although the submarine is approaching a condition of neutral buoyancy. Two things remain yet to be done: 1) to take on additional weight, and 2) to distribute this weight so that fore-and-aft athwartship balance is maintained. This additional weight is added to the *variable ballast tanks* and distributed throughout the variable tanks by the trim system. With the ship in this condition, approximately 55 tons of water must be added to the variable tanks to submerge to a depth where the periscope shears are even with the waterline. The ship is not in a state of neutral buoyancy and is balanced both fore-and-aft and athwartship. At this point, any additional ballast taken on will cause the submarine to submerge; any ballast removed will cause it to rise (Figure 4-1).

However, neutral buoyancy is only a theoretical condition and is very difficult to maintain in practice unless the force of

buoyancy is assisted by some outside force. On the submarine, this assistance is provided by the bow and stern planes and by the propellers. If the trim adjustment is reasonably accurate, the ship will be easily controlled by its planes and speed. To cruise at this depth, the main motors are started. To go to periscope depth, the submarine can plane down with the bow and stern planes. However, to go down in a hurry, it must change from a condition of neutral buoyancy to a condition of negative buoyancy. This is done by flooding the negative tank. The submarine will then be diving 7 tons negative, and must blow negative, therefore, to level off at any given depth, leaving only a water seal in the tank as it approaches the desired depth, thus restoring neutral trim.

If it is desired to surface after returning to periscope depth, the safety tank is blown to restore positive buoyancy, and the bow buoyancy tank is blown to give the ship a rise angle. Should a greater freeboard be desired at the time of surfacing, the main ballast tanks must also be blown. Note that the variable tanks are not blown. These tanks control the trim of the submarine. Therefore, as long as the tanks contain the adjusted weights of water, the ship is in a condition of diving trim.

C. MAIN BALLAST TANKS

4C1. Function and location. The main ballast tanks are water ballast tanks. They are designated as main ballast tanks because they account for the greater percentage of the water ballast

4C2. Description. The main ballast tanks are provided with two to eight flooding openings, located at the lowest point possible on the outer hull. These openings, located in MBT No. 1, MBT

normally carried. They have as their primary function the destroying or restoring of positive buoyancy.

The main ballast tanks (MBT), Figure A-4, are located outside the pressure hull. All A and C tanks are on the starboard side; all B and D tanks are on the port side. Tanks No. 1 and No. 7 extend from port to starboard. All other main ballast tanks are located between the pressure and outer hull and are separated by light athwartship bulkheads.

Nos. 2A, 2B, 2D, 6A, 6B, 6C, 6D and MBT No. 7, are free flooding and are not provided with flood valves. Main ballast tanks No. 2 and No. 6 have, in addition to their primary function of destroying or restoring positive buoyancy, a secondary function of list control.

All main ballast tanks have hydraulically operated vent valves which can be rigged for hand operation. Each tank has a vent riser extending from the top of the tank to the superstructure on the ship's

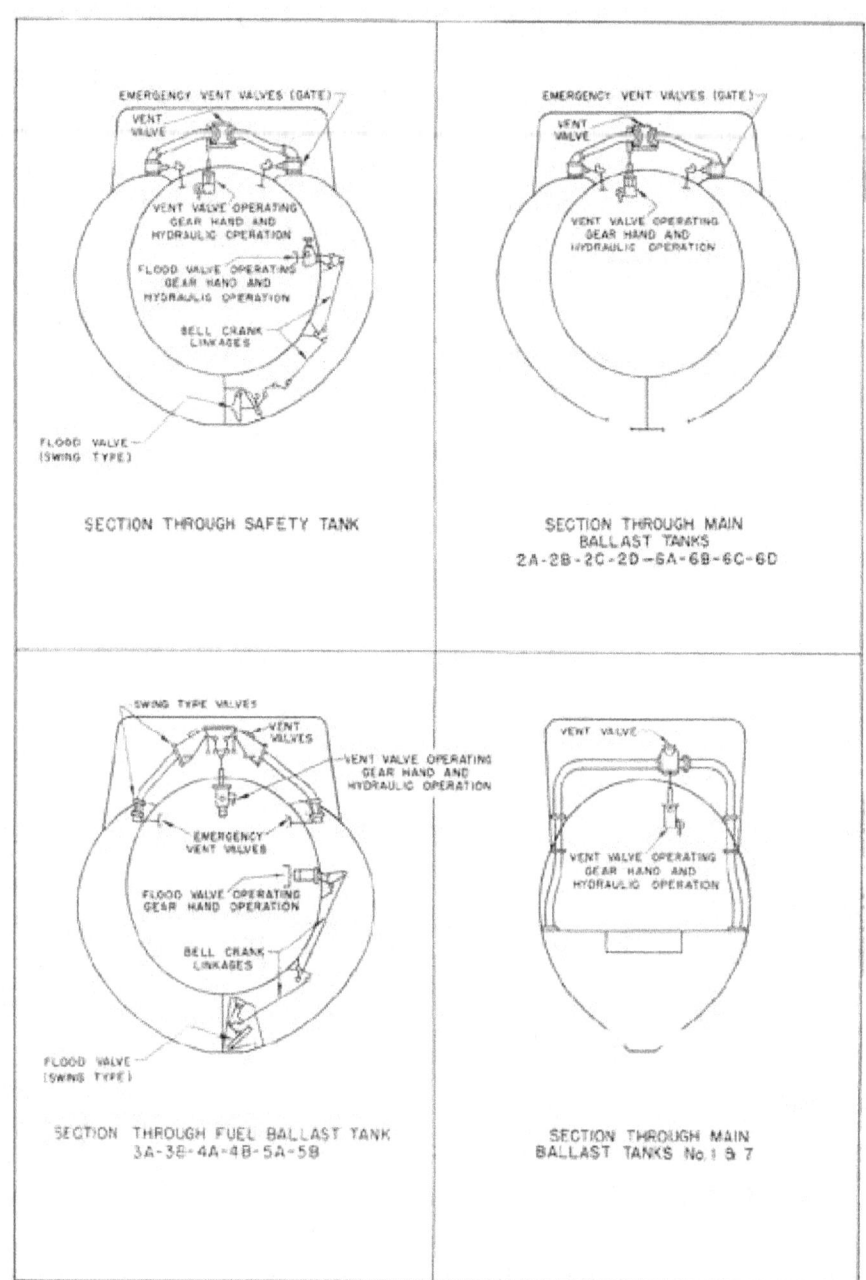

EMERGENCY VENT VALVES (GATE)
VENT VALVE
VENT VALVE OPERATING GEAR HAND AND HYDRAULIC OPERATION
FLOOD VALVE OPERATING GEAR HAND AND HYDRAULIC OPERATION
BELL CRANK LINKAGES
FLOOD VALVE (SWING TYPE)

SECTION THROUGH SAFETY TANK

EMERGENCY VENT VALVES (GATE)
VENT VALVE
VENT VALVE OPERATING GEAR HAND AND HYDRAULIC OPERATION

SECTION THROUGH MAIN BALLAST TANKS
2A-2B-2C-2D—6A-6B-6C-6D

SWING TYPE VALVES
VENT VALVES
VENT VALVE OPERATING GEAR HAND AND HYDRAULIC OPERATION
EMERGENCY VENT VALVES
FLOOD VALVE OPERATING GEAR HAND OPERATION
BELL CRANK LINKAGES
FLOOD VALVE (SWING TYPE)

SECTION THROUGH FUEL BALLAST TANK
3A-3B-4A-4B-5A-5B

VENT VALVE
VENT VALVE OPERATING GEAR HAND AND HYDRAULIC OPERATION

SECTION THROUGH MAIN BALLAST TANKS No. 1 & 7

4-1 Venting and flooding arrangement.

45

centerline. All main ballast tanks, except MBT No. 1 and No. 7, have emergency vent valves located at the tank top, with stems for hand operation extending through and into the pressure hull, to act as an emergency stop valve if the main vent The main ballast tanks are blown with 600-pound air through the 600-pound manifold or with 10-pound air from the low-pressure blower through the 10-pound blow manifold. The low-pressure blower is used only after the ship has

valves or risers are damaged. (See Figure 4-2.)

Sea water is admitted to each ballast tank through flood ports, located in the bottom of the tanks near the keel. They are rectangular in shape.

surfaced.

Each main ballast tank is provided with a salvage air connection which permits blowing the tank from the outside of the hull during salvage operations. Air for such an operation is furnished by a salvage ship through a hose.

D. VARIABLE BALLAST TANKS

4D1. Name and location. There are six variable tanks named and located as follows:

1. Forward trim tank - inside pressure hull

2. Forward WRT tank - inside pressure hull

3. Auxiliary tank No. 1 - outside pressure hull

4. Auxiliary tank No. 2 - outside pressure hull

5. After WRT tank - inside pressure hull

6. After trim tank - inside pressure hull

The general location and shape of each of the variable tanks are shown in FigureA-5.

4D2. Function. The variable ballast tanks are used in conjunction with the trim system to maintain the trim of the submarine. Secondary function of the WRT tanks is to receive water drained from the torpedo tubes and to furnish water for flooding the tubes prior to firing operations.

There are no direct sea connections provided for the variable tanks. All pumping and flooding of these tanks must be done through the trim manifold and the lines of the trim system. The WRT tanks and the trim tanks can be flooded through the torpedo tubes. Blowing and venting of the variable tanks are accomplished by the 225-pound service air system, through the 225-pound manifold, and the torpedo tube blow and vent manifold.

E. SPECIAL BALLAST TANKS

4E1. Safety tank. The primary function of the safety tank is to provide a means for quickly regaining positive buoyancy by blowing the tank when submerged. It follows, then, that the safety tank must be fully flooded when submerged, otherwise it cannot fulfill its primary purpose. For this reason, in the design of the tank

the pressure and outer hulls. (See FigureA-6.) It extends from port to starboard. Since the flooded weight of the safety tank ballast is approximately equal to the weight of water in a flooded conning tower, the safety tank may be blown to compensate for a flooded conning tower.

arrangements, the safety tank has been located so that it has little or no effect on fore-and-aft trim when fully flooded or when blown dry.

The safety tank is located amidships between fuel ballast tanks 3A and 3B and auxiliary ballast tanks No. 1 and No. 2 between

The safety tank is provided with two flood valves, normally operated hydraulically from the control room, with provisions made for hand operation. Indicator lights, operated by a contact at the valves, show whether the safety tank flood valves are opened or shut. These valves open outboard and seat with sea pressure.

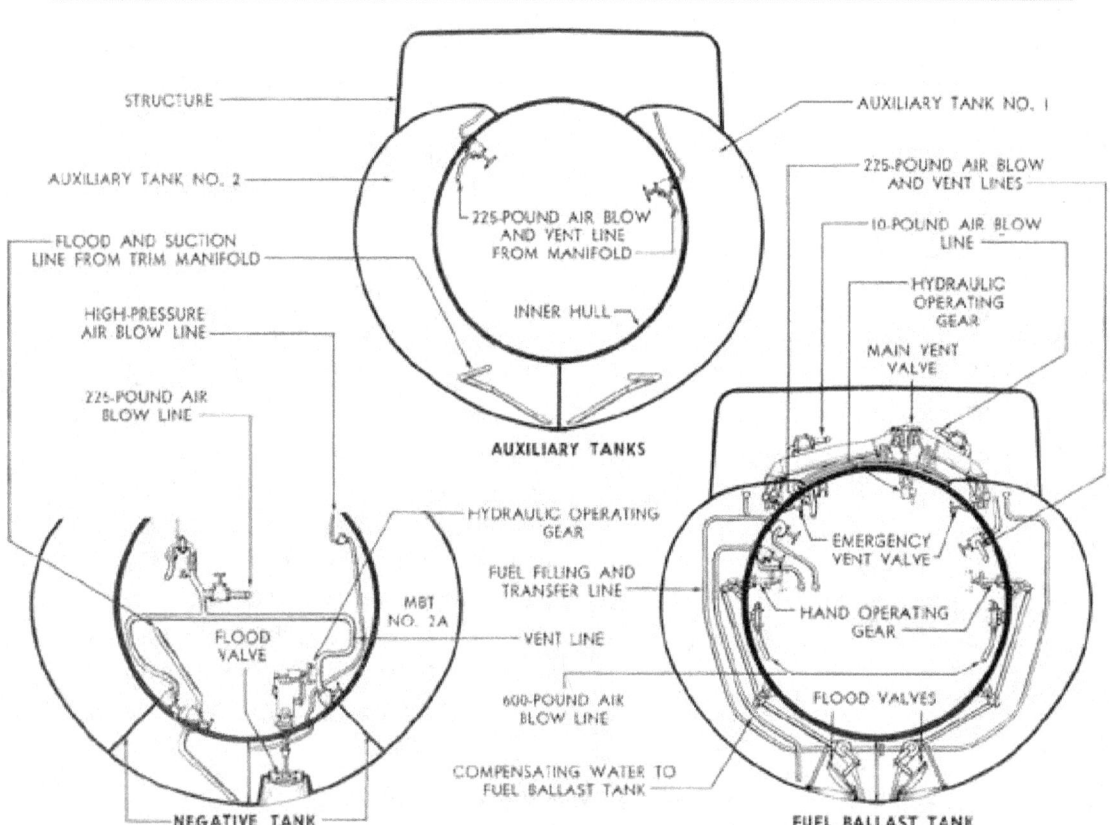

Figure 4-2. Tank connections.

One vent valve, with risers located on both the port and starboard sides, is provided for the safety tank. This vent valve is operated hydraulically from the control room, or locally by hand.

control room, but can be operated locally by hand. This valve opens against sea pressure. The tank vents inboard through a quick-opening manually operated valve located in the control room. (See Figure 4-1.)

The emergency vent valves provided for the safety tank are located port and starboard at the tank top, and are gate type valves, with the gates traveling on the threaded operated stem.

The inboard vents for the safety tank are operated manually from the crew's mess room. These vents located at both the port and starboard sides of the tank connect through a T to one common outlet in the control room. Therefore, when the safety tank is vented inboard, the vented air is bled into the control room proper.

The safety tank is blown by the 3,000-pound air system through the high-pressure distributing manifold in the control room. Like the pressure hull, the safety tank is heavily constructed to withstand full submergence pressure. It is also connected to the trim system and can be used as a variable ballast tank.

4E2. Negative tank. The negative tank, located inside MBT No. 2A and No. 2B, is used to provide negative buoyancy for quick diving.

It is provided with a flood valve which is normally operated hydraulically from the

The tank is blown through the negative tank blow valve on the high-pressure distribution manifold. It can also be blown with the 225-pound system.

The negative tank is built to withstand full submergence pressure and can be used as a variable tank and pumped through the lines of the trimming system.

4E3. Bow buoyancy tank. The bow buoyancy tank is located at the bow of the submarine in the foremost section of the superstructure, as shown in Figure A-6. It is used to correct excessive down angles and to give the ship an up angle during surfacing.

The bow buoyancy tank is free flooding through ports in the superstructure plating along the outside boundary of the tank.

It is provided with two interconnected vent valves which are equipped for hydraulic or hand operation and are controlled by a single operating gear in the forward torpedo room.

The tank outboard vent is protected by a grating in the superstructure deck. The bow buoyancy tank is blown directly with high-pressure air from the 3,000-pound air system.

F. FUEL BALLAST TANKS

4F1. General. There are three fuel ballast tanks, identified as FBT 3A and 3B, FBT 4A and 4B, and FBT 5A and 5B. All A tanks are located on the starboard side; all B tanks are on the port side. The primary function of

ballast tanks only when they fuel is expended.

The fuel ballast tanks are located between the pressure and outer hulls. (See Figures A-4 and A-7.) The A and B

these tanks is to carry ballast, hence they are considered as ballast tanks. However, the secondary function, which is almost as important as their primary function, is to carry reserve fuel oil. In any case, they serve as ballast tanks, since they must be completely full to submerge. With the demand for extended submarine patrols, these tanks are usually filled with fuel oil and become water tanks are connected through limber holes cut in the vertical keel to permit the flow of ballast between tanks.

All fuel ballast tanks have *hydraulically operated* vent valves which can be rigged for hand operation. Each tank has a vent riser extending from the top of the tank in the superstructure on the ship's centerline. All fuel ballast tanks have emergency vent valves located at the tank top, with stems for hand

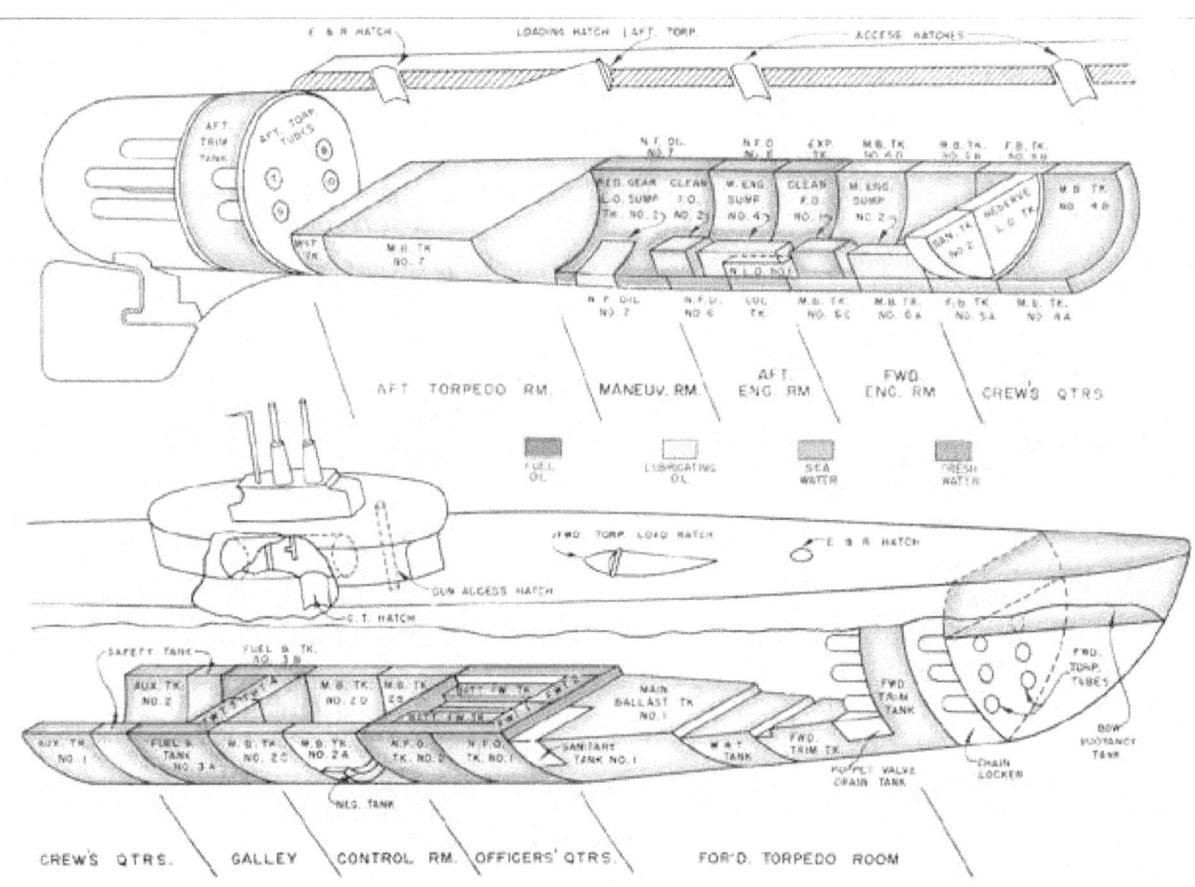

Figure 4-3. Tank arrangement.

operation extending through and into the pressure hull. These valves serve
When the fuel ballast tanks are used as fuel tanks, the flood valves are locked

as stops and enable the tank to be blown if the main vent valves or risers are damaged. (See Figure 4-2.)

Sea water can be admitted to each fuel ballast tank when used as main ballast tanks, through flood valves located in the bottom of the tank near the keel. They are rectangular in shape and open inboard. The fuel ballast tanks are provided with hand-operated flood valves.

shut, the vents disconnected, and a special plate bolted across the vent opening in the superstructure. The fuel ballast tanks are connected to the fuel system, to the compensating water system, and to the 225-pound air system. (See Figure 4-2.)

The fuel ballast tanks, when used as main ballast tanks, are blown with 600-pound air through the 600-pound manifold or with 10-pound air from the low-pressure blower through the 10-pound blow manifold.

G. ADDITIONAL TANKS

4G1. Normal fuel oil tanks. There are four normal fuel oil tanks for the storage of oil for the ship's engines. They are located between the inner and outer hull as shown in Figure A-7. These tanks are known as normal fuel oil tanks (NFOT) Nos. 1, 2, 6 and 7.

Each tank has connections for filling, transferring, and admitting compensation water to replace expended oil, and connections to the 225-pound air system.

Water from the compensating water system is admitted to each fuel tank to compensate for expended fuel oil, or for changes in the volume of fuel oil caused by variations in temperature, thereby keeping the tanks always full of liquid. The fuel tanks are open to sea pressure through the compensating system when submerged.

Try cocks extending into the pressure hull indicate the liquid content of the tank.

volume expands because of temperature variation, and to supply water to the compensating water system. It may also be used to receive bilge water.

Both the expansion and collecting tanks are located between the pressure and outer hull.

4G3. Clean fuel oil tanks. The No. 1 and No. 2 clean fuel oil tanks are located inside the pressure hull; No. 1 is in the after part of the forward engine room, No. 2 is in the after part of the after engine room. Their purpose is to provide main fuel pump suction and to store purified oil.

4G4. The lubricating oil tanks. There are ten lubricating oil tanks, divided into four groups as shown in the following table.

The three normal lubricating oil tanks are used for storage of lubricating oil, as is the reserve lubricating oil tank. The tanks are provided with vents, air connections to the 225-pound air

4G2. Collecting and expansion tanks. The fuel oil collecting tank, located on the starboard side between MBT 6C and NFOT No. 6, is used as a settling tank, separating oil and water in the compensating system to provide a source of oil for the fuel pump. (See Figure A-7.)

The expansion tank, located on the port side between MBT 6D and NFOT No. 6, is used as an overflow tank when the fuel

system, and reducing valves set to deliver air at 13 pounds pressure from the 225-pound service lines. Oil may be blown from any storage tank to any other tank, or discharged overboard, through the lines and manifold of the lubricating system.

The tanks are filled from an outside source by means of a filling connection located on the superstructure deck.

TANK	LOCATION*
Normal lubricating oil group:	
Normal lubricating oil tank No. 1	Forward engine room
Normal lubricating oil tank No. 2	After engine room
Normal lubricating oil tank No. 3	Inside MBT No. 7
Reserve lubricating oil group:	After part, portside after
Reserve lubricating oil tank	battery compartment
Main engine sump group:	
Main engine No. 1 oil sump	Forward engine room
Main engine No. 2 oil sump	Forward engine room
Main engine No. 3 oil sump	After engine room
Main engine No. 4 oil sump	After engine room
Reduction gear sump group:	
Reduction gear oil sump No. 1	Outside pressure hull, inside NFOT No. 7 starboard of centerline
Reduction gear oil sump No. 2	Outside pressure hull, inside NFOT No. 7 port of centerline

*The location of each of these ten oil tanks is shown in Figure A-7.

4G5. Fresh water tanks. There are four fresh water tanks. Nos. 1 and 2 are forward of the forward battery compartment, inside the pressure hull. Tanks Nos. 3 and 4 are located in the

Tanks Nos, 5, 6, 7, and 8-
   After battery space

These are referred to as the forward group (1, 2, 3, and 4), and the after

after part of the control room. (See Figure A-3.) These tanks are used to store the ship's fresh water supply.

There are five emergency fresh water tanks, located one each in the after torpedo room, maneuvering room, control room, and two in the forward torpedo room. The tanks in the forward torpedo room and the one in the after torpedo room can be filled directly from the fresh water system; the other tanks must be filled by portable means.

4G6. Battery water tanks. There are eight battery water tanks used to store the ship's battery water. These tanks are located as follows:

Tanks Nos. 1, 2, 3, and 4-
    Forward battery space

group (5, 6, 7, and 8). (See Figure A-8.)

4G7. Sanitary tanks. There are two sanitary tanks, No. 1 located inside MBT No. 1, and No. 2 located inside the pressure hull in the after starboard end of the after battery compartment. The purpose of these tanks is to collect drain water and refuse from the ship's sanitary system. The No. 1 sanitary tank is connected with the officer's head, while No. 2 sanitary tank is connected with the crew's head in the after battery compartment. (See Figure A-5.)

4G8. Miscellaneous tanks. There are also a number of smaller tanks for special usage. The following table gives the names and locations of these tanks:

51

TANK	LOCATION
1. Hydraulic system supply and vent tank	Control room
2. Three reserve hydraulic oil tanks	Forward torpedo room
3. Hydraulic emergency vent and replenishing tank	After torpedo room
4. Hydraulic emergency vent and replenishing tank	Forward torpedo room
5. Vapor desuperheater tank	Forward engine room
6. Mineral oil tank	Forward torpedo room
7. Mineral oil tank	After torpedo room
8. Torpedo alcohol tank	Forward torpedo room
9. Torpedo oil tank	Forward torpedo room
10. Compressor oil tank	Pump room

52

BUOYANCY AND STABILITY

A. BUOYANCY

5A1. Introduction. Buoyancy is generally understood to be that property of a body that enables it to float on the surface of a liquid or in a fluid. While such a definition is true, it does not fully define the term. Buoyancy, considered in connection with submarines, is the upward force asserted on an immersed

in air and then immersed in water. The aluminum sphere weighs approximately 48 pounds and the cast iron sphere, 136 pounds. If the spheres are lowered into the water, the scale reads 29.1 pounds for the aluminum, and 117 pounds for the cast iron. The differences in weight, 48 - 29.1 = 18.9 and 136 -

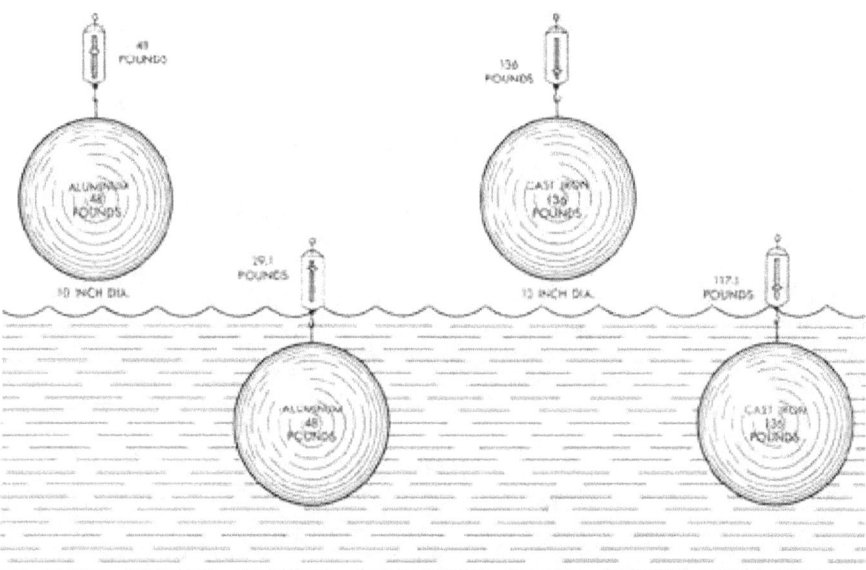

Figure 5-1. Buoyancy depends on volume.

or floating body by the supporting fluid. This conception of the term conveys the idea that *volume*, alone, determines buoyancy, and that the upward force exerted on the immersed or floating body equals the weight of the fluid which it displaces. This idea is illustrated by the diagrams in Figure 5-1.

A sphere of aluminum and one of cast iron, each 10 inches in diameter, are weighed

117.1 = 18.9, are the same, showing that the buoyancy, or upward force of the displaced water, is the same in both cases and is independent of the weight of the immersed body.

The buoyancy of a submarine is also dependent on the volume of the displaced water and it is controlled by varying the volume of displacement as illustrated in Figure 5-2.

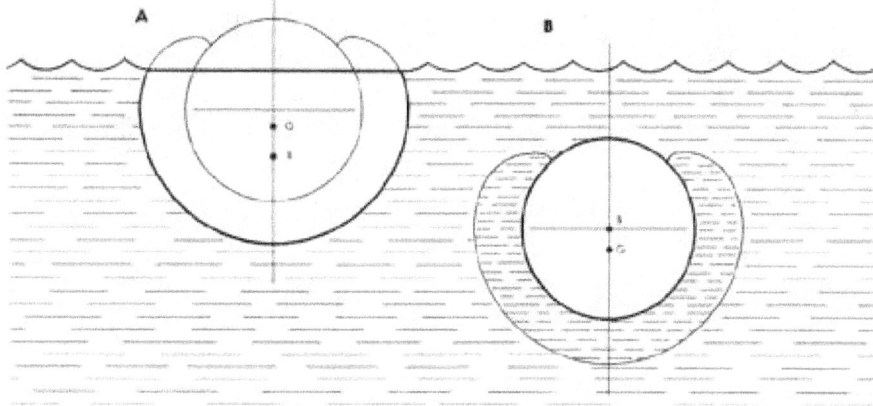

Figure 5-2. Volume of displacement is changed.

Diagram A, Figure 5-2, represents a submarine on the surface. Its main ballast tanks are filled with air. The displacement water, represented by the area within the heavy line, equals the weight of the submarine.

Diagram B, Figure 5-2, represents a submerged submarine. Water has been admitted to the main ballast tanks, expelling the air. The displaced water is now represented by the area within the heavy circle. The over-all weight of the submarine is not changed, but the submarine may be submerged because the volume of displaced water has been reduced and the weight of the displaced water is now the same as or less than the weight of the submarine.

5A2. Center of buoyancy. The *center of buoyancy* is the center of gravity of the displaced water. It lies at the geometric center of volume of the displaced water. The center of buoyancy should not be confused with the center of gravity of the immersed, or floating, body. These two centers are indicated as *B* and *G,* respectively, on the sketches in Figure 5-2.

5A3. States of buoyancy. By definition, buoyancy is the upward force exerted on a floating, or

weight of the body and the weight of the displaced fluid. In the case of submarines, the displaced fluid is sea water. Three *states of buoyancy* are considered: 1) positive buoyancy, 2) neutral buoyancy, and 3) negative buoyancy.

1. *Positive buoyancy* exists when the weight of the body is less than the weight of an equal volume of the displaced fluid.

2. *Neutral buoyancy* exists when the weight of the body is equal to the weight of an equal volume of the displaced fluid. A body in this state remains suspended, neither rising nor sinking, unless acted upon by an outside force.

While this condition might be attained in a laboratory, it is doubtful that it is ever obtained exactly in a submarine. Nevertheless, the condition is approached and any discrepancy is counteracted by the diving planes; the ship is then considered to be in a state of neutral buoyancy.

3. *Negative buoyancy* exists when the weight of the body is greater than the weight of an equal volume of the displaced fluid and the body sinks.

Theoretically, a submarine is designed

immersed, body and is independent of the weight of the body. The state of buoyancy refers to the ratio between the

with its main ballast tanks of such volume that when they are flooded, the ship is in the state of neutral buoyancy. Negative buoyancy is gained by flooding the negative tank.

54

B. STABILITY

5B1. Stability. Stability is that property of a body that causes it, when disturbed from a condition of equilibrium, to develop forces, or moments, that tend to restore the body to its original condition. Because stability is a state of equilibrium, this term should be defined.

5B2. Equilibrium. Equilibrium is a state of balance between opposing forces and may exist in three states: (1) stable, 2) neutral, and 3) unstable.

1. Stable equilibrium is that property of a body that causes it, when disturbed

A cone lying on its side may be rolled on its surface and will remain in its displaced position. A cone may be balanced on its point and remain in equilibrium but, when disturbed, will increase its displacement.

The two conditions, buoyancy and stability, are so closely related and interdependent when considered in connection with submarines that they must be discussed together.

All floating bodies, including both surface ships and submarines, are subject to the same natural forces, and these forces

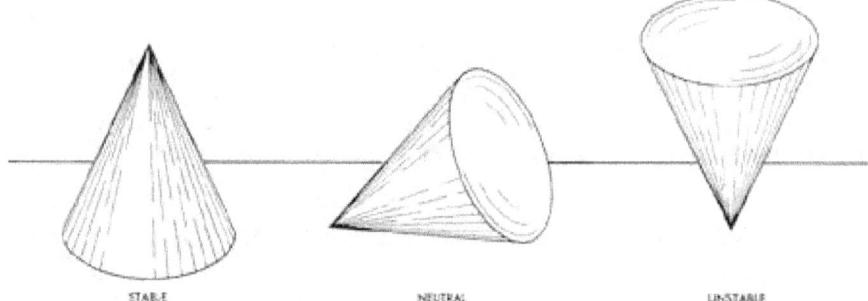

STABLE   NEUTRAL   UNSTABLE

Figure 5-3. States of equilibrium

from a condition of equilibrium, to develop forces, or moments, that tend to restore it to its original condition. When a floating body is in stable equilibrium, its center of gravity and its center of buoyancy are in the same vertical line.

2. Neutral equilibrium exists when a

in all cases follow the same physical laws. There is a difference, however, between the stability of surface ships and the stability of submarines. Because submarines are special cases of floating bodies, their stability requires a special application of these laws.

Another term, *metacenter,* needs to be

body remains in its displaced position.

3. Unstable equilibrium exists when a body tends to continue movement after a slight displacement.

These three states are illustrated in Figure 5-3.

A cone resting on its base may be tipped in any direction, within limits, and will return to is original position when released.

understood before proceeding with the discussion of stability.

5B3. Metacenter. Metacenter is the point of intersection of a vertical line through the center of buoyancy of a floating body and a vertical line through the new center of buoyancy, as shown in the diagrams in Figure 5-4.

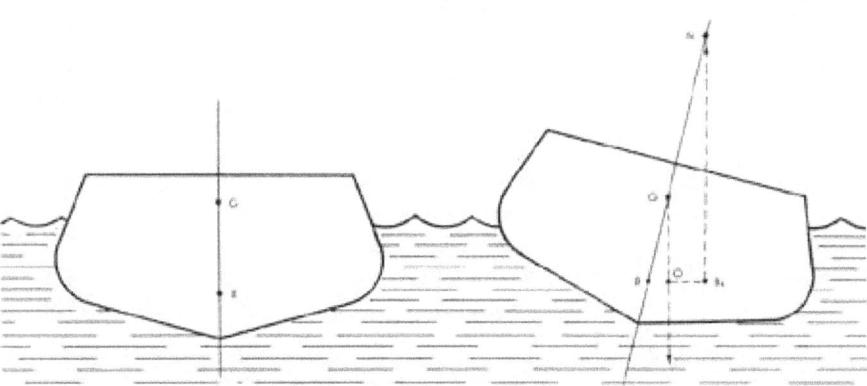

Figure 5-4. The metacenter.

When a vessel is tipped as shown, the center of buoyancy moves from *B* to *B1*, because the volume of displaced water at the left of *G* has been decreased while the volume of displaced water to the right is increased. The center of buoyancy, being at the center of gravity of the displaced water, moves to point *B1*, and a vertical line through this point passes *G* and intersects the original vertical at *M*. The distance *GM* is known as the *metacentric height*. This illustrates the fundamental law of stability. When *M* is above *G*, the metacentric height is positive

and the vessel is stable because a moment arm, *OB1*, has been set up which tends to return the vessel to its original position. It is obvious that if *M* is located below *G*, the moment arm would tend to increase the inclination. In this case the metacentric height is negative and the vessel would be unstable.

When on the surface, a submarine presents much the same problem in stability as a surface ship. However, some differences are apparent as may be seen in the diagrams in Figure 5-5.

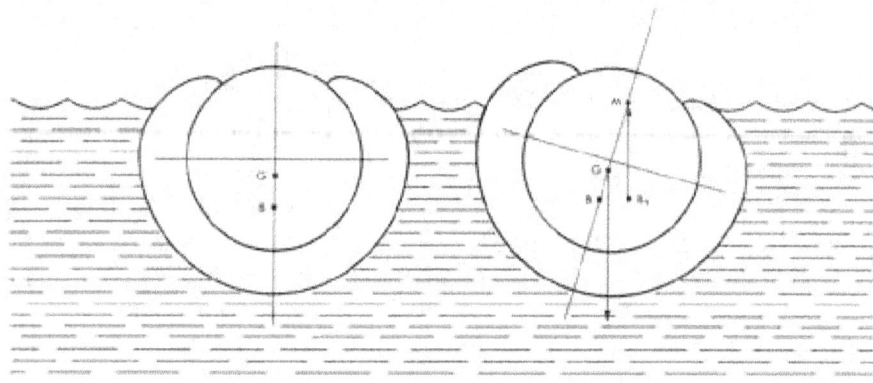

Figure 5-5. A submarine on the surface.

Figure 5-6. Change of center of buoyancy and metacenter during submergence.

The three points, *B, G,* and *M,* are much closer together than is the case with surface ships. When a submarine is submerged, these significant points are arranged much differently.

The center of gravity of the submarine, *G,* remains fixed slightly below the centerline of the boat while *B* and *M* approach each other, *B* rising and passing *G,* until at complete submergence *B* and *M* are at a common point. These changes are shown diagrammatically in Figure 5-6.

On the surface the three points, *B, M,* and *G,* are in the same relative positions as for surface ships. As the ballast tanks fill, the displacement becomes less with the consequent rising of *B* and lowering of *M.* There

is a point during submergence or surfacing when *B* coincides with *G* and *GM* becomes zero or perhaps a negative quantity. During a normal dive, this point is passed so quickly that there is no time for the boat to take a list. When the ballast tanks are fully flooded, *B* rises to the normal center of buoyancy of the pressure hull, and stability is regained with *G* below *B.*

Just why these centers change so radically may be made more readily apparent by an illustration with rectangular sections. The diagrams in Figure 5-7 represent a rectangular closed chamber, so weighted at *G* that it will sink in water. The area surrounding it at the sides and bottom represents air chambers.

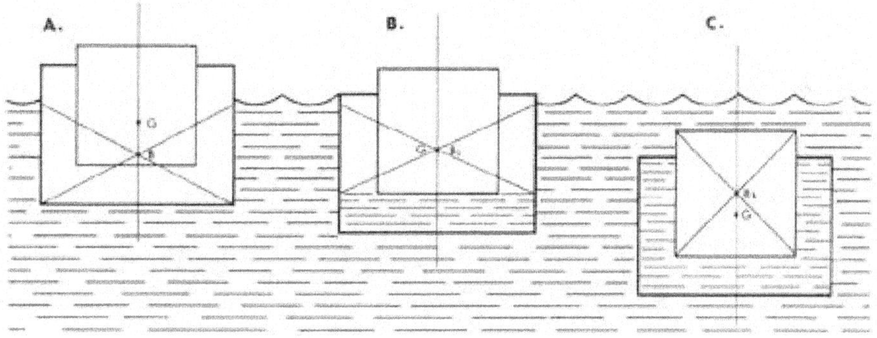

Figure 5-7. The center of buoyancy shifts.

At A, the vessel is floating with all water excluded from the tank surrounding the chamber. The center of gravity is at *G* and the center of buoyancy, *B,* is found by intersecting diagonals of the displacement.

At B, water has been admitted to the lower section of the tank. Using the diagonals as before, it is seen that the center of buoyancy, *B,* is now coincident with *G* and the unit is unstable.

At C, the surrounding tank is flooded and the unit is submerged. The center of buoyancy is at *B2*, the intersection of the diagonals of the displaced water. The unit is stable, the center of buoyancy and the center of gravity are in the same vertical line. Any rotational movement about the center of buoyancy *B2* immediately sets up a restoring moment arm.

When surfacing, with the water ballast being ejected comparatively slowly by the

shape below the waterline all affect stability.

It is an axiom that high freeboard and flare assure good righting arms and increase stability, and low freeboard and "tumble home," or inward slope, give small righting arms and less stability. The diagrams in Figure 5-8 show why this is true.

Diagram A represents a cylindrical vessel with its center of gravity at the center of the body and so weighted that it floats on its centerline. Its center of buoyancy is at the center of gravity of the displaced water.

It is at once apparent that this vessel is not in stable equilibrium. *G* and *B* will remain in the same position regardless of rotation of the body. As no righting arms are set up, the vessel will not return to its original position.

Diagram B represents a vessel of equal volume and the same waterline. Its center

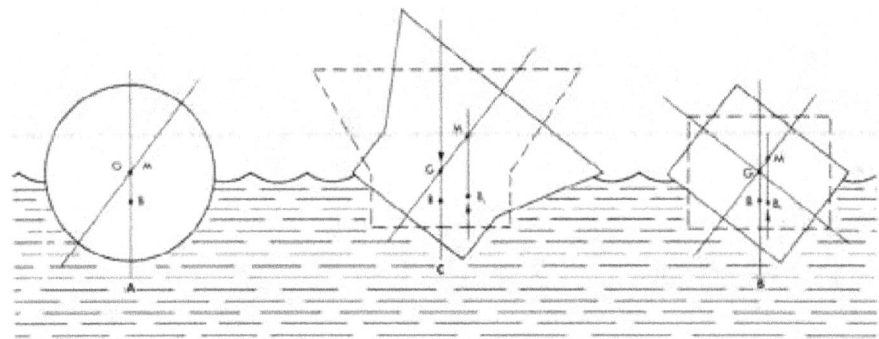

Figure 5-8. The effect of shape and freeboard on stability.

low-pressure blowers, *GM* may become negative and a list may occur. As a corrective measure, if a list should occur, certain main ballast tanks are provided with separate low-pressure blow lines for the port and starboard sections. Lever-operated, list control valves are installed so that air to the tanks on the high side may be restricted and more air delivered to the low side.

5B4. Transverse stability. The stability of any vessel on the surface depends upon two things: 1) the position of the center of gravity, and 2) the shape of the vessel. The shape above the waterline, the freeboard, and the

of gravity is at the center of volume, and the center of buoyancy is at the center of gravity of the displaced water. When this vessel is inclined about its center of gravity, the effect of change of shape is noticeable. The volume of displaced water at the left of *G* is decreased and the displacement at the right is increased. The center of buoyancy moves to the right, the metacenter, *M*, is above *G*, and the force coupled B_1G, tends to right the vessel.

In diagram C the vessel is flared from the waterline and its freeboard increased. When this vessel is inclined, the added

displacement of the flare is added to that resulting from the shape of the underwater section, and the center of buoyancy shifts about four times as far, raising the metacenter and providing a stronger righting arm.

The submarine has the worst possible shape, little freeboard and extreme tumble home. For this reason, every effort is made to keep the center of gravity as low as possible. The storage batteries, weighing approximately 1 ton per cell, and all heavy machinery are set as low as possible, but the

than that resulting from the shorter transverse axis; consequently, the center of buoyancy moves a greater distance. In the ship illustrated, the movement of the center of buoyancy to the right is approximately 23 feet, giving a surface metacentric height of 370 feet. Because of the shorter transverse axis, the transverse metacentric height is only 1 1/2 feet.

When a submarine submerges, however, the water plane disappears and the metacenter comes down to the center of buoyancy. This is because the forces on a superstructure, deck equipment, and conning tower total a considerably high weight. Because of the difficulty of getting the normal center of gravity low enough, submarines usually carry lead ballast along the keel.

submerged body act as if the body is suspended from its center of buoyancy. Being submerged, the volume of displaced water on each side of the center of buoyancy remains constant, regardless of the angular displacement of the axis. As the center of rotation

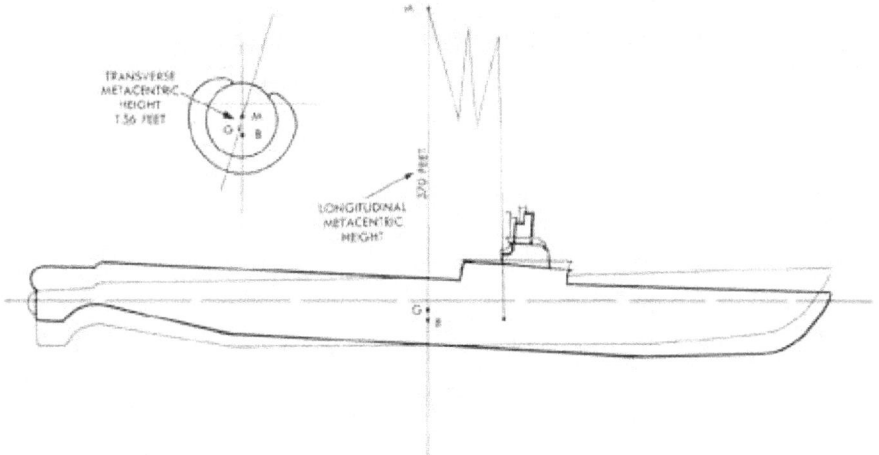

Figure 5-9. Stability increases with length of waterline.

5B5. Longitudinal stability. The longitudinal stability of a submarine is much greater than the transverse stability. Stability in both cases depends on the relative positions of the metacenter and the center of gravity but, in this case, the metacenter is calculated with respect to the longitudinal axis.

Figure 5-9 shows why a slight angular displacement of water resulting from a slight angle of the longitudinal axis is much greater

of a submerged body is at its center of buoyancy, vertical lines through this center, for any position of the body, always intersect at the same point. Thus, for a submerged body, the metacenter and center of buoyancy are coincident. This agrees with the definition of metacenter given on page 55. Therefore, the longitudinal *GM* and the transverse *GM* are the same for a submerged submarine except for the effect of free surface.

5B6. Free surface. Free surface refers to the surface of ballast water in a partially filled tank in which water is free to move and to assume its normal surface. The adverse effect of free surface in a ballast tank may be visualized from the diagrams in Figure 5-10.

submerged. The fuel ballast tanks are connected with the sea and the fuel is forced out at the top. Thus they are always filled either with oil or water or some proportion of each.

During submergence, when the ballast tanks are being flooded and again when they are being blown for surfacing, there

Free surfaces affect longitudinal stability more than transverse stability because of the greater moment arms involved.

Diagram A represents a tube, with closed ends, partially filled with water.

It is suspended at *B*, the exact center of its length. The center of gravity of the water is at *G*. *B* and *G* are on the same vertical line and the tube is in equilibrium. If

is a period during which free surface exists in all main ballast tanks. To reduce the effect of this free surface the ballast space is divided by transverse bulkheads into a number of separate tanks. The effect of this division of the longitudinal ballast space is illustrated in diagram C, Figure 5-10.

Partitions are indicated in the tube, each section has its own center of gravity,

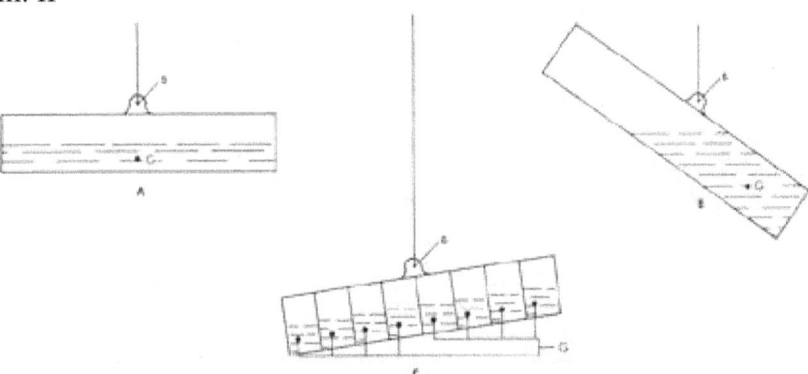

Figure 5-10. Effect of free surface on stability.

the tube is disturbed even slightly, the free surface permits the water to flow to the low end and as shown in diagram B. Any movement of the water toward either end moves the center of gravity and sets up a moment arm, increasing both the inclination and the moment arm. This continues until the water is in one end of the tube and *G* is again on the same vertical line with *B*.

If the tube is filled with water, eliminating all free surface, there can be no movement of the water, and the unit acts as a solid and remains in stable equilibrium.

Submarines are designed to eliminate, as far as possible, all free surfaces. The main ballast tanks are proportioned so that they are completely filled when the vessel is

and free surface exists in all sections. As the tube is inclined, the water shifts as before but to a limited extent, and the cumulative result of the shifting of the individual centers of gravity is negligible.

In a submarine this result of a momentary free surface in the tanks is counteracted by the diving planes. However, water collecting in a flooded compartment will seriously affect both longitudinal and transverse stability.

5B7. Addition of permanent weight. The effect of adding weight to a submarine is serious, not only because it makes the vessel heavy, but also because of the consequent reduction in stability. The addition of weight to a surface ship causes it to sink a little lower in the water, increasing displacement and usually, stability. If the weight added is below the center of gravity, stability is further increased.

With the submarine, conditions are different, for, in order to be in readiness for submerging with the main ballast tanks empty, she must always float at the same waterline. To meet this requirement, the weight must be constant, as it is not possible to alter the capacity of the main ballast tanks or the buoyancy of the hull without structural changes. Auxiliary tanks are provided for the usual variations in weight of fuel,

stores, crew, and so forth, but the addition of permanent weight would require the removal of an equal amount of permanent ballast. The added weight, if above the center of gravity, raises the original center of gravity; removal of ballast raises it still more, resulting in a reduction of the normally short righting arm and reducing stability. The addition of deck armament or any deck load should be carefully considered as to its effect on the center of gravity.

Disregard of the laws of stability will render the submarine less seaworthy and may invite disaster.

61

6
ENGINEERING PLANT

A. TYPE OF DRIVE

6A1. General Electric drive set. The later fleet type submarines use the diesel-electric drive for main propulsion.

The primary source of power is four main diesel engines. The engines are used to drive four main generators which in turn supply the electric power to drive the four main propulsion motors. Each of the two propellers is driven by two main propulsion motors through a production gear. (See Figure A-9.)

When submerged, the diesel and

c. *Two reduction gears*: each is single reduction with two pinions driven by two main propulsion motors and a slow speed gear connection to the propeller shaft. Input is 1,300 rpm, output is 280 rpm. This reduction gear is not used on the latest type submarines which are equipped with slow-speed main motors.

d. *One auxiliary generator*: direct current, rated at 300 kw at 1,200 rpm, and driven by a directly connected diesel engine. It is used to supply current to the auxiliary power system and for battery charging.

generators are not used. Power for the motors is supplied by two sets of storage batteries. These batteries are charged by the auxiliary and main generators during surface operations.

The electric propulsion machinery installed in each 1,500-ton class submarine includes the following (See Figure 6-1.):

a. *Four direct-current generators*: each rated 1,100 kw, 415 volts, 750 rpm. Each generator is cooled by a surface air cooler and generates power for driving the propulsion motors and charging the storage batteries. Each generator is driven by a directly connected diesel engine.

b. *Four direct-current propulsion motors*: each rated 1,375 hp, 415 volts, 1,300 rpm. Two of these motors are used to drive each propeller, through a reduction gear, at 280 rpm. Each motor is equipped with an air cooler.

e. *The electric propulsion control*: includes equipment required for the control and operation of the four main motors when being supplied with power generated by one or more of the main generators or by the storage batteries. The equipment also provides for controlling the charging of the storage batteries by the auxiliary or main generators. Main power cables, capable of carrying the heavy currents involved, connect the storage batteries with the various main generators and motors.

In these circuits, there is control equipment consisting of switches, resistance units, and protective devices designed to permit flexibility of control.

Each unit of the electrical propulsion machinery listed above, the accessory equipment, and the diesel engines, are discussed in the following sections of this chapter.

B. ENGINES

Main engines. Four main engines are employed to drive the main generators. There are two in each engine room. (See figure A-9.)

Each engine is of the 2-cycle diesel type;

the two on the port side are arranged for left-hand rotation, those on the starboard, for right-hand rotation. The general characteristics of the General Motors engine are shown in the following table:

62

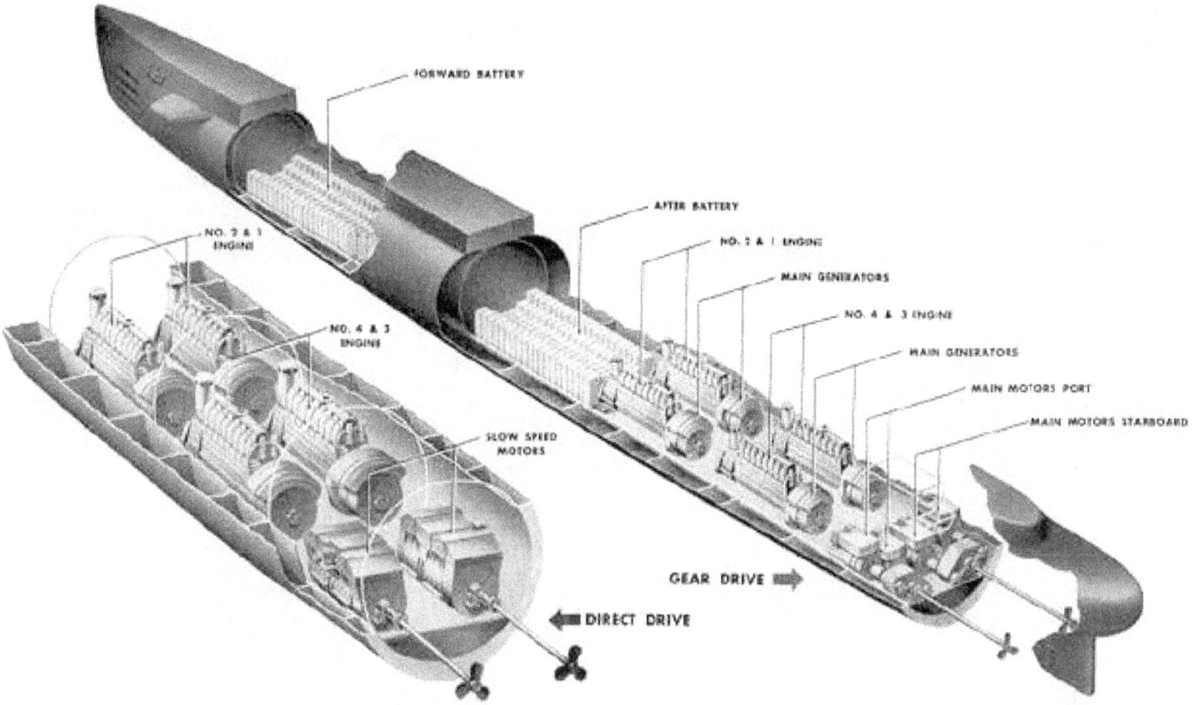

Figure 6-1. General arrangement of main propulsion equipment.

63

Part	Characteristics
Model number	16-278A
Brake horsepower	1,600
Rated engine speed	750 rpm
Cylinder arrangement	V-type
Number of cylinders	16
Bore and stroke	8 3/4 X 10 1/2
Starting system	Air starting

The diesel engine differs from the internal combustion engine in that it has no spark plugs or ignition system, and relies on heat of compression to ignite the fuel which is forced into the cylinders under pressure and atomized.

one-piece cut-joint type. The connecting rods are made from an alloy steel forging.

The crankshaft is of heat-treated alloy steel with eight crank throws spaced 45 degrees apart, and with nine bearing surfaces.

The camshafts, which control the action of the valves, and the accessory drive shaft are driven by gears from the crankshaft. The accessory drive furnishes power to the blower and the water pumps. The camshaft drives the oil pump, the tachometer, and the engine speed governor.

A diesel engine burns a mixture of fuel oil and air. A blower is provided on

The two types of main engines used are the Fairbanks-Morse 38D81/8 and the G. M. 16-278A. A description of the G. M. Engine follows.

The cylinder block is the main structural part of the engine. It is fabricated from forgings and steel plates welded together, combining strength with light weight. The upper and lower decks of each cylinder bank are bored to receive the cylinder liners. The space between the decks and the cylinder banks forms the scavenging air chamber. Removable hand-hold covers close the openings in the sides of the cylinder blocks and provide access to the interior of the engine for inspection or repair.

The cylinder liner is made of alloy cast iron, accurately bored and finished. It is removable and can be replaced when worn.

The engine cylinders are fitted with individual alloy cast iron cylinder heads. Each head is fitted with four exhaust valves, the unit injector, the rocker lever assembly, an engine overspeed injector lock, the cylinder test and safety valves, and the air starter check valve.

The pistons are made of alloy cast iron. The bored holes in the piston pin hubs are fitted with bronze bushings. Each piston is fitted with seven cast iron rings, of which five are compression rings, and two are oil control rings. These rings are of the conventional

each engine to furnish air to the cylinders, and to remove burnt gases from the cylinders. This function is known as *scavenging*. The scavenging blower consists of a pair of rotors revolving together in a closely fitted housing and furnishes a constant, uniform supply of air at the rate of 5,630 cubic feet per minute. The blower is mounted directly on the engine at the forward end and is driven by the accessory drive shaft.

Fuel for each engine is drawn from the clean oil tank by the fuel oil pump, forced through a filter, and delivered to the injectors. The pump is mounted on the blower end of the engine and is driven directly by one of the camshafts.

The governor, which controls the engine speed, is of the self-contained hydraulic type. It is provided with a power mechanism to regulate the fuel injectors, thus increasing or decreasing speed. The governor is set to allow a maximum engine speed of 802 rpm.

The engine is cooled by a fresh water cooling system. The fresh water is circulated by the fresh water pump, which is of the centrifugal type, and is mounted on the blower end of the engine.

The purpose of the sea water cooling system is to conserve fresh water by cooling it after its passage through the engine, thereby

permitting its constant reuse for

engine lubricating oil system and to a

cooling purposes. The sea water pump, adjacent to the fresh water pump, is also of the centrifugal type.

The engine is started by air, which is furnished by the high pressure (3,000-pound) air system. Air is admitted through the starting air valve at 500 pounds' pressure to the engine cylinder, causing the engine to start. When the engine begins to run under its own power the air is shut off.

Each engine is fitted with a flexible coupling, providing direct connection to the generator it drives.

6B2. Lubrication. Each main engine is provided with a pressure oil system for lubrication.

The lubricating oil pressure pump on each engine draws oil from the sump tank through a check valve and forces it through an oil strainer and cooler. From the oil cooler, the lubricating oil is delivered to the

branch line that supplies the generator bearings.

The generator bearing scavenging oil pump draws oil from the generator bearings and returns it to the engine oil pan.

The lubricating oil enters the engine at a connection on the control side of the camshaft drive housing. A regulating and relief valve adjusts the pressure of the oil as it enters the engine.

From the engine inlet connection, the oil flows to the main lubricating oil manifold, where it is distributed to the main bearings, piston bearings, connecting rod bearings, camshaft drive gear and bearings, and the valve assemblies. The excess oil drains back to the oil sump. Oil is supplied by another branch to the blower gears, bearings, and rotors.

The lubricating oil pressure pump, mounted on the camshaft drive housing cover, is a positive displacement helical spur gear type pump.

C. ELECTRICAL EQUIPMENT

6C1. Description of main generators. The main generators are two-wire, direct-current, separately excited, shunt-wound, multi-pole, totally enclosed, and self-ventilated machines. The armature shafts are supported on a bearing at each end, except in the Elliott machine which employs a single bearing at the commutator end. The bearings are force lubricated by the oil supply from the main engine lubricating system.

employed on the other types. The main generators are rated at approximately 2,650 amperes at 415 volts and 1,100 kw.

6C2. Cooling systems. The cooling systems of the various machines operate on the same principle. The hot air is cooled by forcing it through water-cooled cores. The older Allis-Chalmers machines, however, do not employ the duct work used on the other type machines. The cooling unit on these

The maximum speed of a main generator is dependent upon the type of main engine. Maximum speed with a G. M. Engine as a prime mover is 750 rpm; with a Fairbanks-Morse engine, 720 rpm. Direct flexible coupling to the engine is accomplished through the flanged end of the generator armature shaft.

With the exception of the cooling units on Allis-Chalmers machines, the construction of all main and auxiliary generators is similar. However, in the latest Allis-Chalmers machines, the encircling cooling arrangement has been replaced by a system similar to that

generators fits the contour of the machine and is made in two sections, each half section covering one-fourth of the outer surface of the generator. Water tubes are set in grooves on the inner surface of the shell to absorb the heat from the circulating air.

The other type machines have the water tubes mounted in cores, similar to an automobile radiator. This assembly is located in the air ducts of the cooling system through which the air passes.

Circulating of air is by means of the ventilating fan attached to the armature

shafts of all machines. Air is delivered from the cooler into the commutator and housing. It is then drawn through the field coils and through the commutator ends, under the commutator into the armature, and then through ventilating ducts in the armature core.

6C3. Description of the auxiliary generator. The 300-kw direct-current auxiliary generator is a two-wire shunt compensated (G.E. and Elliott) or differential compound (Allis-Chalmers) machine. The generator is self-excited, but the switching is arranged so that separate excitation may be obtained from the battery. The rating limits of the machines are approximately 345 volts at 870 amperes and 300 kw at 1,200 rpm.

The generator is connected to the auxiliary diesel engine through a semi-rigid coupling. The commutator end of

Various combinations of armatures in series or in parallel, including all four motors in series for dead slow operation, may be obtained for either surface or submerged operation through the main control cubicle.

Motor speed control is accomplished by controlling the *generator speed and shunt field,* thus varying the voltage supplied during surface operation, and the *motor shunt field* when submerged. Reverse operation is accomplished by reversing the direction of the flow of current in the motor armature circuit.

Main motors used in a gear drive installation are classed as high-speed motors and are each rated for continuous duty at approximately 1,375 hp, 415 volts, 2,600 amperes, 1,300 rpm.

6C5. Lubrication. Oil under pressure is supplied to the motor bearings by a gear-driven lubricating oil pump which

the armature shaft is supported on a sleeve bearing which is force lubricated from the engine lubricating system. The opposite end of the shaft is carried by the engine bearing. The generator armature thrust is taken by thrust faces on the ends of the sleeve bearing.

In construction, auxiliary generators differ only in minor detail from the main generators. They are produced by the same manufacturers and with the exception of size, weight, and number of some of the components, auxiliary and main generators are identical.

6C4. Description of main motors. The main motors are of the two-wire, direct-current, compensated compound type with shunt, series, and commutating field windings. Separate excitation for the shut field is provided by the excitation bus which receives power from either battery.

The motors are totally enclosed, water-tight below the field frame, split and water-proof above. Cooling is accomplished by a fan attached to the armature shaft which circulates the air through cores cooled by circulating water.

Each end of the armature shaft is supported on a split sleeve bearing. The bearings are lubricated from the oil supply in the reduction gear units.

is attached to the reduction gear units of each pair of motors. However, when the propeller shaft speed is operating at a slow speed, a standby pump is placed in operation and supplies sufficient oil pressure for both reduction gears and main motor bearings. Oil-catching grooves and return drains into the housing prevent leakage of oil along the shaft into the windings. The air chamber between the bearing and the interior of the motor serves to prevent formation of a vacuum around the shaft and permits drainage of any possible oil leakage before it reaches the interior of the motor. A safety overflow is provided in the housing oil reservoir to prevent possible flooding of the winding, if the drain should become clogged. After passing through the bearing, the oil passes out of the housing through a sight flow and returns to the lubricating oil sump. When the flow of oil at the sight flow glass appears to be appreciably reduced, or, if the oil pressure falls below 5 psi, the standby pump must be placed in operation. The standby system is also used to force the lubricant to the bearings before starting the motors after a shutdown period.

6C6. Cooling system. The main motor cooling units are similar to the main generator

units with one exception. The Allis-Chalmers cooling units on the older main motors are made in three sections which cover approximately 90 percent of the outer surface of the motor

power ranging from 20 hp to 2,700 hp per propeller shaft at speeds ranging from approximately 67 rpm to 282 rpm.

For submerged operation, using various

frame. The remaining surface is covered with a dummy section to secure the necessary clearance for the motor arrangement in the motor room. This arrangement is such that each motor has its cooler sections placed on different portions of its outer surface.

6C7. Description of the double armature propulsion motor. On the latest type submarines, main motors and reduction gears have been replaced by two 2,700-hp double armature motors, directly connected to the propeller shafts, one to the starboard, the other to the part shaft.

The motors are of the two-wire, direct-current, compounded, compensated type with shunt and series field windings and commutating poles. Separate excitation for shunt fields is provided by the excitation bus which receives power directly from the battery in the control cubicle. The motors are totally enclosed with a water tube air cooler mounted crosswise over the motor frame. Mechanical air filters are located in the air ducts between the coolers and vent blower. When the motors are operating in the *SLOW* position, neither cooling air nor circulating water is required.

The motor frame is split at an angle of approximately 11 degrees from the horizontal center line to permit easy removal of the armature. The motor is watertight below this joint and waterproof above.

The armature is mounted on a hollow forged steel shaft which is flanged at the after end for coupling to the propeller shaft. Each end of the shaft has a bearing journal for a force-

combinations of armatures and taking power from the batteries, the motors will develop power ranging from 30 hp to 1,719 hp per propeller shaft, and will give a speed range from 42 to 219 rpm.

6C8. Main control equipment. Fundamentally, the construction of main propulsion control equipment produced by General Electric (See Figure 6-2), Westinghouse, and Cutler-Hammer is similar. Individual components may vary somewhat in design; their locations and method of installation in the assembly may differ; cables and conduits will be found routed differently; but, each assembly as a whole performs the same function and is operated in a similar manner.

6C9. Split type main propulsion control equipment. (See Figure 6-3.) The split type control equipment is installed on some of the later type submarines on which double armature, slow speed, directly connected propulsion motors are used. This equipment performs the same functions as the standard control cubicle, and with minor exceptions is operated in the same manner.

The two halves of the control panel are essentially the same. Each half is mounted in a steel frame that is joined to form a single unit and shock mounted to the hull. The starboard control panel consists of the generator levers for the No. 1 and No. 3 generators, starting with reversing levers for the starboard motor, a bus selector, and a forward battery lever. The port control panel consists of the generator levers for the No. 2 and No. 4 generators, starting and reversing levers for the port motor, a bus selector,

lubricated, split sleeve bearing, mounted in a pedestal, separate from the frame. In addition to the radial bearing, the forward end of the shaft is fitted with a collar for a Kingsbury thrust bearing which takes the propeller and motor thrust load.

For surface operation, using the various combinations of armatures and taking power from the main generators, the motors develop

and after battery lever.

6C10. Functions. The control equipment performs the following functions:

1. Start, stop, reverse, and regulate the speed of the main motors for both surface and submerged operation.

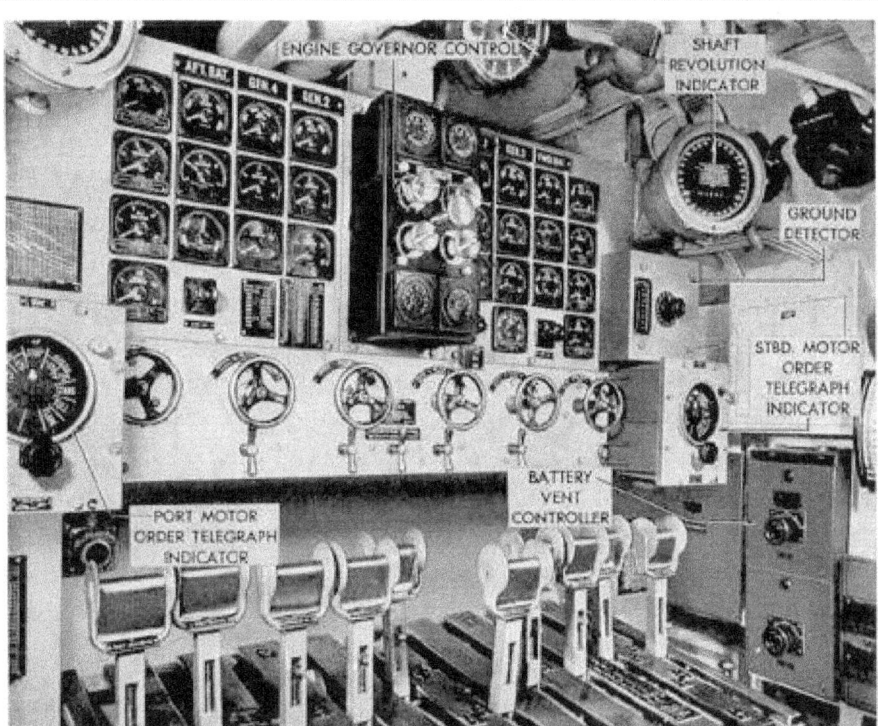

Figure 6-2. General Electric main propulsion control cubicle.

2. Provide for series, parallel, or series-parallel connection of the motor armatures.

3. Provide for uniform speed control of the main motors throughout the entire range of propeller speed from about 42 to 219 rpm submerged to about 280 rpm on the surface.

independent of each other except for a common excitation bus.

7. Provide for operation ahead on one propeller shaft and astern on the other at any speed within the designed operating range.

8. Provide, by means of shore connections, for charging the main

4. Provide for operating the main motors from one or both main storage batteries and from any combination of the main generators.

5. Provide for charging one or both storage batteries with main generators individually or in combination. Main generators not being used for battery charging may be used for propulsion power.

6. Provide for driving the starboard motors from the starboard generators and the port motors from the port generators entirely

battery from shore or tender.

6C11. Simplified circuit description. The main control cubicle circuit consists essentially of two buses, the motor bus and the battery bus, to which the main power units are connected by means of their associated contactors in order to provide the various operating combinations. The motor bus is the one to which the main motors are connected for any of the running conditions by means

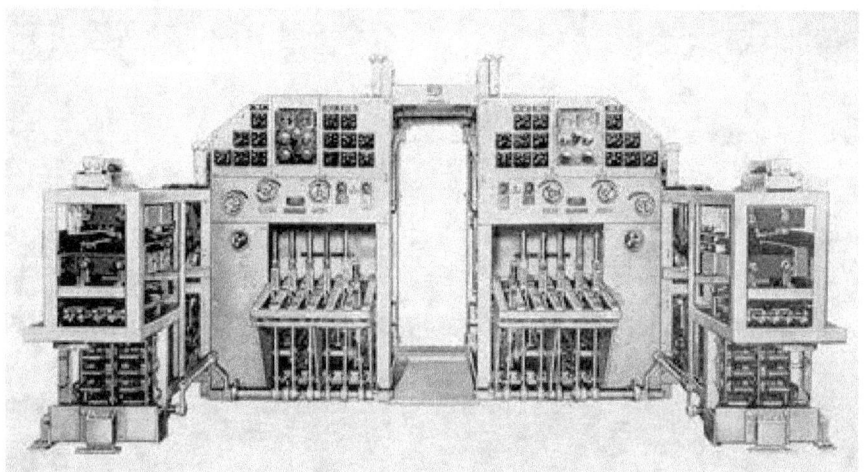

Figure 6-3. Split type of main propulsion control cubicle.

of their starting contactors.

The motor bus can be split for operation of the motors on each side independently of the other sides (BUS TIE OPEN), closed for parallel operation of both motor groups (BUS TIE CLOSED), connected to the battery bus for battery operating of the main motors (BATTERY BUS), and lastly, for series operation of all motors the positive side of one motor

a. Port and starboard motor reversing switches.
b. Port and starboard motor starting contactors.
c. Bus selector switches.

3. One forward contactor group comprising:

a. Port and starboard main generator contactors.
b. Forward and aft battery contactors.

bus can be cross-connected to the negative side of the other motor bus, so that by proper closing of the motor contactors, all four motors can be placed in series for slow-speed operation on the battery bus (SLOW).

Either or both batteries are connected to the battery bus by closing their respective contactors which, in turn, are controlled by one operating lever.

6C12. Principal parts. The principal parts of the equipment are as follows:

1. One main propulsion control panel and operating bench with necessary instruments, rheostats, and operating levers.

2. One aft contactor group comprising:

c. Motor bus tie contactors.

All parts are mounted in a number of steel frames which in turn are joined to form a single unit. The assembly is supported on rubber shock mounts that are welded to the hull.

6C13. Operating levers. There are 10 levers for manual operation of the contactors in the various switch groups. These levers are provided with lock latches and are connected mechanically to the contactor camshafts through a series of bell cranks and rods. The purpose of the levers is as follows:

a. *Two reverse levers.* These levers are

used to change direction of rotation of the main motors by reversing the current flow through the armature. One lever is for the two starboard motors and the other for the two port motors. Each lever has three positions, OFF, AHEAD, and ASTERN.

b. *Two starter levers.* Each of the starter levers for the two port and starboard motors has a STOP position and five operating positions, SER. 1, SER. 2, SER. 3, PAR 1, PAR 2. The starter lever is used for cutting in resistance in series with the armature, thus keeping the starting current down to a minimum. As the motor picks up speed, the resistance can be cut out of the circuit when the armature is at running speed and the current reaches a normal value, putting it across the

any one or all generators on the battery bus for charging the batteries, or any one or all generators on the motor buses for propulsion. An extra mechanical latch on each lever prevents accidental movement from the OFF position.

d. *One battery selector lever.* This lever has an OFF position and three operating positions, AFT BAT, FWD BAT, and BOTH BAT. Placing the lever in the AFT BAT position will place the after battery on the battery bus. Placing it on the FWD BAT position will place the forward battery on the battery bus. In the BOTH BAT position, both batteries are in parallel with each other and on the battery bus. The battery bus is a common connection which is supplied with current from either one or both batteries and which in turn supplies line voltage. The starter levers have three series positions and two parallel positions. The motors are always in series with each other when the starters are in any of the series positions, the voltage of the line being divided between each of the motors. When the starters are in either parallel position, the motors are in parallel, each motor receiving the full line voltage. The SER. 3 and PAR. 2 positions are the only running positions of the starter levers.

c. *Four generator levers.* One lever is provided for each of the four main generators. The levers have one OFF position and two operating positions, MOTOR BUS and BAT. BUS.

Note. On Westinghouse controls, the operating positions are GEN. BUS and BAT. BUS.

The function of these levers is to place current to the motor bus for motor propulsion when the bus selector is in battery position. In addition, any or all of the generators may be placed on this bus to charge either one or both batteries as desired. When the battery bus is used only for charging, it is necessary to have only the battery selector and the charging generator on the battery bus; the bus selector can be in the OFF position.

e. *One bus selector lever.* The bus selector has five positions, BUS TIE CLOSED, BUS TIE OPEN, OFF, BAT BUS, and SLOW. The function of this lever is to keep the port and starboard motor buses in an open or closed position, to connect the battery bus with the motor bus, and to close the necessary contactors to operate all motors in series.

D. AUXILIARY EQUIPMENT

6D1. Description. Approximately 40 auxiliary motors of various capacities are located throughout the ship for operation of compressors, blowers, pumps, and other miscellaneous equipment. Current for operation of these motors is supplied by the auxiliary generator, the main batteries, or a combination of both, through two auxiliary distribution switchboards. The forward distribution switchboard, connected to the forward battery, feeds all auxiliary machines in and forward of the control room, while the after distribution switchboard, powered by the after battery or the auxiliary generator, feeds all auxiliary machines aft of the control room. A bus-tie circuit connects the two switchboards, making it possible to feed one switchboard from the other in the event of an emergency.

During normal operation, the bus-tie circuit is left open and the power for both switchboards is taken from the

6D4. Main storage batteries. Each ship has two main storage batteries consisting of two groups of 126 cells

batteries with the auxiliary generator often floating on the line. The batteries are thus paralleled through the battery selector on the main control panel. With the circuit so connected, the auxiliary generator will contribute current not used by the auxiliary load toward charging the batteries. This circuit arrangement is also used when the auxiliary generator is secured.

6D2. Auxiliary motors. Auxiliary motors are direct-current motors designed to operate on a voltage ranging from 175 volts to 345 volts. Their horsepower rating type of winding, and other data are stamped on the name plate attached to each motor. Auxiliary motor frames are enclosed to provide protection against dripping water and are vented to permit the escape of hot air which is forced out by a fan attached to the armature shaft. Magnetic disc brakes are used on motors that must stop after the current is shut off.

6D3. Motor-generator sets. The following are two types of motor-generator sets:

a. *Lighting motor-generator sets.* These machines are used on some ships to deliver current for operation of the lighting system as well as for the I.C. motor-generator sets that require a lower voltage than that delivered directly by the battery or auxiliary generator. The 175-345 d.c. motor receives its power from the battery or auxiliary generator and through a common shaft driver the 120-volt generator. It is controlled by a speed regulator.

Note. On some ships lighting motor-

each. The forward battery is installed below decks in the wardroom country and the after battery is located in the crew's space.

6D5. Battery installations. The installation in each battery tank consists of 6 fore and aft rows of 21 cells each. The two center rows are on one level. The rows alongside the center are slightly higher and the outboard rows are the highest. Above the two central rows of cells are installed panels of hard rubber that serve as a working deck or flat.

All cells are connected in series by means of intercell connectors while end cells in each row of batteries are connected by means of end cell connectors.

Both positive and negative forward battery disconnect switches are manually operated from a station at the after end of the forward battery room.

The after battery disconnect switches, also manually operated, are located near the after end of the crew's quarters. These disconnect switches are used only in an emergency to isolate the battery.

6D6. Battery ventilation. Each battery is fitted with an exhaust ventilating system in order to remove battery gases. The intakes for the air required to operate this system are located at opposite ends of each compartment. The free air in the compartment is drawn through the filling vent connection of each cell. The cells are connected by soft rubber nipples to exhaust headers of hard rubber which extend fore and aft for each row of cells.

generator sets have been replaced by lighting feeder voltage regulators.

b. *I.C. motor-generator sets.* I.C. motor-generator sets are d.c.-a.c. machines equipped with speed and voltage regulators to produce a 60-cycle current for certain interior communication systems. The d.c. motor receives its power from the lighting motor-generator on ships so equipped, or directly from the battery or auxiliary generator on ships that do not use lighting motor-generator sets.

The headers are in two sections and are connected to cross headers which unite in a common exhaust duct. The exhaust duct from each battery is led up to and through the deck to the inlets of two fans which are mounted on the hull overhead in the respective battery rooms.

Each of these four fans is rated at 500 cubic feet per minute at 2,700-revolutions per

minute. Each fan is independently driven; the motor is controlled from the maneuvering room. The motors used on late type submarines are rated at 1.25 hp (continuous), 2,780 rpm, 175-345 volts, 5.0 amperes.

Starting and speed regulation are accomplished by armature resistance. A fused tumbler switch for each motor is mounted in a separate case and connects both the armature circuit and the field circuit to the supply lines. Each armature circuit includes armature resistance, sections of which may be short-circuited by a 20-point dial switch to provide speed control. *In regulating the battery ventilation by means of armature resistance and adjustments, care must be taken (when two blowers are being used for one battery) to set the pointers on both rheostat knobs to approximately the same point.* The power supply is obtained through a fused switch marked BATTERY VENTILATION on the after distribution switchboard in the maneuvering room.

6D9. Lighting system. The lighting system is composed of the ship's service lighting system and the port and starboard emergency lighting systems. Each is a separate distribution system.

Power for the ship's service lighting system on late type submarines is obtained from the batteries through two lighting feeder voltage regulators and a lighting distribution switchboard. On earlier ships, power for this system was supplied by lighting motor-generator sets.

On some ships, a battery selector switch has been incorporated in the lighting distribution switchboard and permits selection of either battery or the shore connection as the source of power.

The feeders from the lighting distribution switchboard run the length of the ship on both sides and serve all regular lighting circuits through fused feeder distribution boxes. Final distribution to lighting fixtures and low-current outlets is through standard lighting distribution boxes with switches

A damper is provided in the duct between the inlets of the two fans to allow the fans to be operated singly or together. *When a single fan is used, the damper must be set in order to close the inlet to the idle fan, thereby preventing free circulation of air through both fans.* Each pair of fans exhausts into the ship's exhaust system.

6D7. Air flow indicator. An indication of the quantity of air passing through the ventilation system is given in the maneuvering room by means of two air flow indicators. Each of these indicators is provided with a scale marked in cubic feet per minute.

6D8. Air flow through individual cells. The flow of air through cells of each battery compartment is equalized by means of adjusting regulators which are installed as an internal part of each filling vent cylinder. Proper adjustment of these regulators has been determined and set at a navy yard and must not be altered by ship's personnel.

Note. Do not permit the soft rubber nipples to attain a twisted position. A twisted or partially collapsed nipple will materially affect the ventilation to the respective cell or cells.

and midget fuses for each outgoing circuit.

The starboard emergency lighting system is directly powered through two cutout switches which are connected to the positive and negative end cell terminal connections of the forward battery. These switches are connected to 13 lighting units, a circuit to the auxiliary gyro, and the forward and after marker buoy circuits. A branch junction box provides a connection to the gyrocompass control panel for the alarm system.

The port emergency lighting system is directly powered through cutout switches connected to the after battery. The arrangement of this system is similar to that of the starboard emergency system except for the location of the circuits and the lack of a gyrocompass alarm connection.

Each lighting unit consists of *two* 115-volt lights, a protective resistor, and a snap switch, all connected in series, as they *always operate directly on full battery voltage.*

E. CIRCUITS AND SWITCHBOARDS

6E1. Circuits. The interior communication systems in a submarine provide the means of maintaining contact, transmitting

- General announcing system: circuit 1MC
- Submarine control announcing system: circuit 7MC

orders, and relaying indications of the conditions of machinery to other parts of the ship.

The majority of the systems are electrical and automatic in operation. Some, such as the motor order telegraph system, are manually operated, but include an electrical circuit for the actual transmission of the order to another part of the ship. Some of the systems operate on alternating current; still others, such as the tachometer and sound-powered telephone systems, operate on self-generated current.

The I.C. systems of a modern fleet type submarine usually consist of about 24 circuits. With few exceptions, they are supplied with power through the I.C. switchboard located in the control room.

Following is a list of important I.C. circuits and their circuit designations:

- Telephone call system: circuit E
- Engine governor control and tachometer system: circuit EG
- Battle telephone systems: circuits JA and XJA
- Engine order control system: circuit 3MB
- Dead reckoning tracer system: circuit TL
- Collision alarm system: circuit CA
- General alarm system: circuit G
- Diving alarm system: circuit GD
- Low-pressure lubricating oil and high-temperature water alarm system: circuit EC

- Rudder angle indicator system (selsyn): circuit N
- Bow and stern plane angle indicator system: circuits NB and NS
- Auxiliary bow and stern plane angle indicator system: circuits XNB and XNS
- Main ballast indicator system: circuit TP
- Hull opening indicator system: circuit TR
- Underwater log system: circuit Y
- Bow plane rigging indicator
- Target designation system: circuit GT

In addition, the following circuits, which are not part of the interior communication systems, are supplied through switches on the I.C. switchboard:

- Circuit Ga-1: torpedo data computer
- Circuit 17Ga-1: torpedo data computer
- Circuit 6Pa: torpedo firing
- Circuit 6R: torpedo ready lights
- Circuit GT: target designation system

6E2. Systems requiring alternating current. The following systems require alternating current for operation:

- Self-synchronous operated motor order telephone and indicator systems (for main propulsion orders)
- Self-synchronous operated diving plane angle indicators (bow and stern planes)
- Self-synchronous operated rudder angle indicator system
- Hull opening and main ballast

- Shaft revolution indicator systems: circuit K
- Gyrocompass system: circuit LC
- Auxiliary gyrocompass system: XLC
- Motor order telegraph system: circuits 1MB and 2MB
- Marker buoy system: circuit BT

- tank indicator systems
- Hydrogen detector
- Lubricating oil (low-pressure) and circulating water (high temperature) alarm systems

- General announcing systems (alarm signals and voice communication)
- Torpedo data computer
- Self-synchronous operated underwater log system
- Self-synchronous operated propeller shaft revolution indicator system
- Target designation system
- Engine governor control system (direct current on latest classes)
- Torpedo data computer (Ga-1)

6E3. Systems requiring direct current. The following systems require direct current for operation:

- Marker buoy system
- Engine order indicator system (alternating current on older submarines)
- Searchlight
- Auxiliary gyrocompass
- Torpedo data computer (17Ga-1)
- Torpedo firing
- Torpedo ready lights
- Engine governor control system
- Auxiliary bow and stern plane

electrical devices that assist in solving the fire control problem and firing the torpedo tubes. The devices that perform these functions are the torpedo data computer, the gyro-angle indicator regulators, and the torpedo ready and firing light systems.

The torpedo data computer is located in the conning tower and is energized by circuits Ga-1 (115-volt alternating current) and 17Ga-1 (115-volt direct current).

The gyro-angle indicator regulators for automatically setting the gyro angles on the torpedoes in the tubes are located at the forward and after tube nests and are controlled by separate fire control circuits from the torpedo data computer. The regulators are supplied with 115-volt direct current through circuits 17Ga-3 and 17Ga-4.

In addition to the above, the following circuits are provided:

- An underwater log repeater circuit to the torpedo data computer
- A gyrocompass repeater circuit

angle indicating systems
- Resistance thermometer systems
- Gyrocompass system

6E4. Interior communication switchboard. The I.C. switchboard is usually located on the starboard side of the control room. The switchboards on the latest type submarines are equipped with snap switches and dead front fuses with blown fuse indication mounted directly below or on either side of each switch requiring fuses. Earlier type submarines used knife switches with fuses mounted immediately below each switch.

a. *Source of power.* The alternating-current power supply to this switchboard is obtained from the I.C. motor-generators which are 250-volt d.c. motors and 120-volt a.c. generators.

6E5. Torpedo fire control system. The torpedo fire control system employs several

to the torpedo data computer

6E6. Torpedo ready light, torpedo firing, and battle order systems. The torpedo ready light, torpedo firing, and battle order systems provide a means of informing the fire control party when the tube is ready to fire, or directing the tube crew to stand by a tube to fire, or firing the torpedoes remotely from the conning tower and simultaneously indicating to the tube crew by means of an audible and visual signal that the tube has been fired, and lastly of indicating that the tube has been fired, by a visual signal in the conning tower. It also indicates to the fire control party in the conning tower, by means of a visual signal, that the gyro-angle indicator regulators are matched.

6E7. Torpedo ready light and battle order system. A forward and after transmitter in the conning tower is used to transmit torpedo orders to an indicator at each tube nest. When power is turned on in the conning tower, the READY AT TUBE and the

STANDBY pilot lights on the corresponding indicator in the torpedo room are lighted. When the gyro retraction spindle switch contacts are closed, an amber GYRO SPINDLE IN light in the indicator in the conning tower for that tube is lighted. When the tube interlock switch contacts are closed, the amber READY AT TUBE light for that tube in the indicator in the torpedo room is lighted. When the operator turns the indicator switch for that tube, it lights a green READY light for the tube in the transmitter in the conning tower. When the gyro

corresponding green STANDBY light for that tube in the indicator in the torpedo room is lighted. Upon pressing the firing contact maker, the tube is fired through operation of the pilot valve solenoid. Simultaneously a red FIRE light in the indicator is lighted and a buzzer is operated at the tube nest.

6E8. The torpedo firing system. The torpedo firing system (Circuit 6PA) is energized from the 120-volt direct-current bus on the I.C. switchboard.

Separate fixed and portable contact

mechanism of the tube nest is matched upon closing the manual contact of the gyro setting mechanism contact maker, a red angle SET light in the transmitter in the conning tower is lighted. When a particular tube standby switch in the conning tower is turned to STANDBY at the transmitter, the

makers (firing keys), for independently controlling the forward and after groups of firing solenoids, are located at the torpedo ready light and firing panels in the conning tower. A key, mounted on the gyro-angle regulator indicators, operates a light on this panel to show that the regulator is matched.

7
VENTILATION SYSTEM

A. VENTILATION ARRANGEMENT

7A1. General. The ventilation arrangement consists of four systems: 1) the engine induction, 2) the ship's supply, 3) the ship's exhaust, and 4) the battery exhaust. The ship's supply and ship's exhaust systems are cross-connected as are also the ship's exhaust and battery system. (See Figure A-10.)

7A2. Engine induction. Air to the engine is supplied by natural induction through a 36-inch diameter ventilation stack and outboard valve in a compartment in the after end of the conning tower fairwater. The bottom of this stack has three branches, one 16-inch diameter branch for ship's supply and two 22-inch diameter branches, one port and one starboard. All the branches are located in the superstructure. The port pipe runs aft to a hull valve in the top of the inner shell, near the middle of the forward machinery compartment. The starboard pipe runs aft to a hull valve in the top of the inner shell near the middle of the after machinery compartment. These hull valves permit

combination engine induction and ship's supply outboard valve (Figure 7-1) located in the after end of the conning tower fairwater. A 16-inch diameter pipe in the superstructure runs from this stack to a hull valve in the top of the inner shell near the forward end of the forward machinery compartment. The 4000 cubic feet per minute ship's supply blower, which receives its air through this hull valve, attached fitting and inlet pipe, is just aft of the former and discharges into a splitter fitting directing air into both forward and after supply mains. This splitter fitting has a fixed damper that has been adjusted to divide the flow forward and aft in the right proportions. The supply mains run to the forward and after compartments of the vessel and are equipped with valves, branches, and louvers for adequate air distribution to all necessary spaces.

7A4. Ventilation blowers. Ventilation blowers are employed to move air within the hull supply, hull exhaust, and battery exhaust systems. Each of the blowers is

passage of air directly to their respective machinery compartments. Just forward of the after hull valve a 22 x 22 x 16-inch lateral bypass dampcr fitting is inserted in the induction pipe, the 16-inch outlet of which discharges into a 16-inch pipe, running aft to an auxiliary hull valve in the top of the inner shell in the maneuvering room, aft of the maneuvering control stand. This air, when thus bypassed, is for the purpose of creating lower temperatures in the vicinity of the control stand and eventually circulates forward again for consumption by the engines in the machinery compartments through the door or exhaust bulkhead valves in the forward maneuvering room bulkhead.

7A3. Ship's supply. During normal surface operation, ship's air is supplied through the

driven by an individual direct-drive electric motor.

Figure 7-2 is an illustration of a typical ventilation blower of the type used on the submarine.

The hull supply blower has a capacity of 4000 cubic feet per minute. It is powered by an electric motor rated at 2 to 5 horsepower.

The hull exhaust blower has a capacity of 2560 cubic fee per minute at 1750 rpm. Its motor is rated at 2 to 4 horsepower.

Each of the two battery exhaust systems has two blowers. These blowers are rated at 500 cubic feet per minute at 2780 rpm. Power is supplied to each blower by a 1 1/4-horsepower electric motor. A damper, placed between the inlet ports of the two

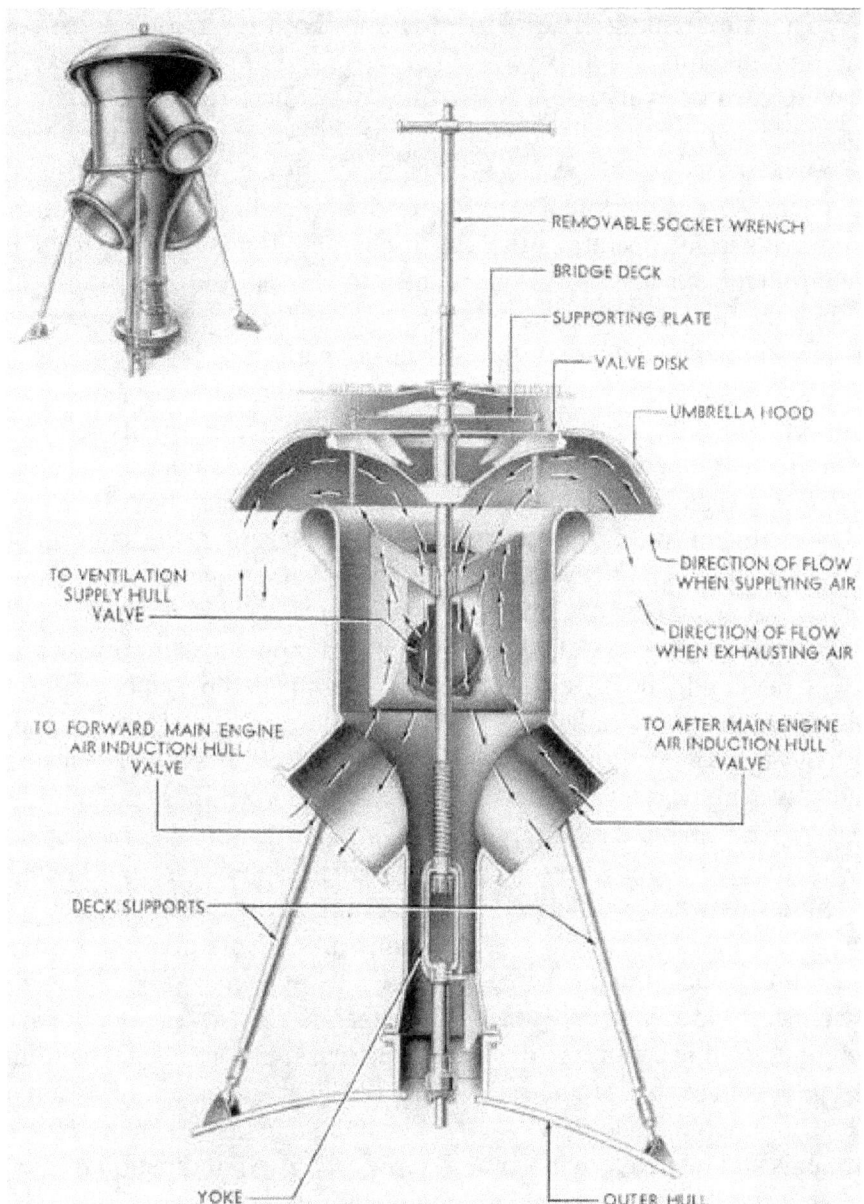

REMOVABLE SOCKET WRENCH

BRIDGE DECK

SUPPORTING PLATE

VALVE DISK

UMBRELLA HOOD

DIRECTION OF FLOW
WHEN SUPPLYING AIR

DIRECTION OF FLOW
WHEN EXHAUSTING AIR

TO VENTILATION
SUPPLY HULL
VALVE

TO FORWARD MAIN ENGINE
AIR INDUCTION HULL
VALVE

TO AFTER MAIN ENGINE
AIR INDUCTION HULL
VALVE

DECK SUPPORTS

YOKE

OUTER HULL

Figure 7-1. Engine induction and ship's supply outboard valve.

77

blowers of each battery exhaust system, permits the use of one or both blowers.

7A5. Ship's exhaust. The 2500-cubic feet per minute ship's exhaust blower is in the forward end of the forward machinery compartment and receives to a single fitting in the platform deck above the batteries. (See Figure 7-3.) To the top of this fitting is attached a riser, leading up to a damper fitting connected to the inlets of two battery exhaust blowers. Either or both blowers may be operated at any one time for each battery. The forward set is located in the

its air from a main, running forward to the after end of the forward torpedo room, having adequate valves, branches, and louvers from all necessary spaces forward of the forward machinery compartments. During normal surface operation, the discharge from the exhaust blower is directed into the forward machinery compartment through a damper-controlled louver in a fitting adjacent to the blower. When the ship is at the dock or laying to, with engines not running, this air, plus that discharged into all compartments aft of the crew's quarters by the ship's supply system, may find its way overboard via open hatches or engine induction hull valves, whichever may be most desirable under existing weather conditions. Any doors or exhaust bulkhead valves in the after bulkheads, necessary to permit natural egress of exhaust air from the boat by the above-mentioned means, should be left open. When the submarine is cruising on the surface, the discharge from the exhaust blowers is consumed by the engines. In this condition all airtight dampers in supply branches should be shut in both machinery compartments and maneuvering room. If air is being supplied to the after torpedo room, exhaust bulkhead valves must be left open in both forward and after maneuvering room bulkheads so that exhaust air may get back to machinery compartments for consumption by the engines.

7A6. Battery exhaust. Each of the two battery tanks normally receives its supply from the compartment above via two battery intakes, one near each end of the room. In an emergency, these may be closed off by gas-tight

overhead of the chief petty officers' quarters and the after set is overhead of the crew's mess room. Each pair of blowers discharges into its own common fitting, having a partial damper for flow regulation. These fittings discharge into the ship's exhaust main near the battery blowers.

During periods of battery exhaust, gases have high hydrogen content (see Section 7A7), a damper, normally left entirely or partly open to the inlet of the ship's supply blower, may be fully closed, thereby diverting all exhaust air (which contains battery gases) overboard through the normal ship's supply outboard piping system. In such a position, the damper uncovers a screened opening permitting inboard supply to the ship's supply blower from the forward machinery compartment.

An orifice plate is inserted in the battery exhaust riser from each battery compartment, and pressure fittings from each side of each orifice plate are piped to separate air flow meters in the maneuvering room. The type of meter used is an arrangement of differential pressure gage known as the *Hays Air Flow Meter.* An indication of the quantity of air flow indicators. Each of these indicators is provided by a scale marked in cubic feet per minute.

The operation of the battery ventilation system covering fan speeds, controller settings, and meter readings is detailed in the *Bureau of Ships Manual,* Chapters 88 and 62.

All piping and fittings in the superstructure are designed to resist sea pressure at the maximum depth of

covers stowed near the intakes. In each battery tank, air is sucked through each cell by a network of hard rubber piping, eventually consolidating into one pipe connecting

submergence.

Three conditions under which it may be necessary or desirable to recirculate the air inside the ship rather than take in air from the outside as described for normal

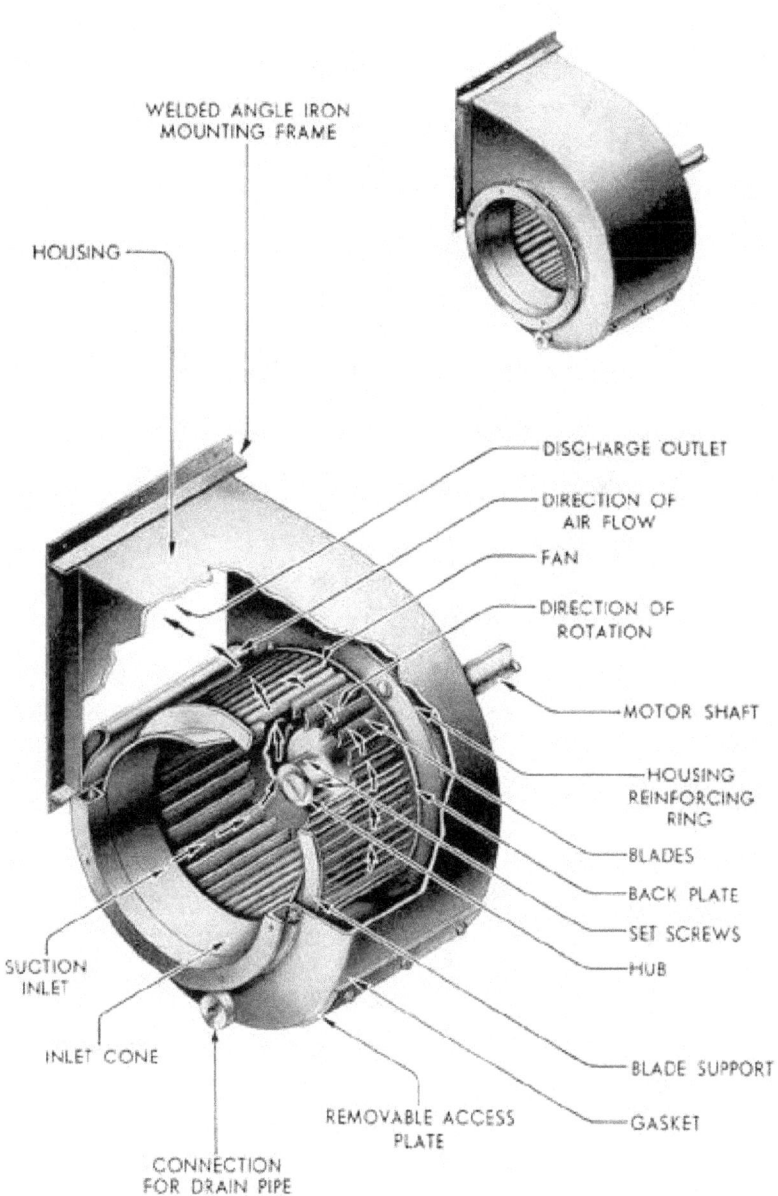

Figure 7-2. Ventilation blower.

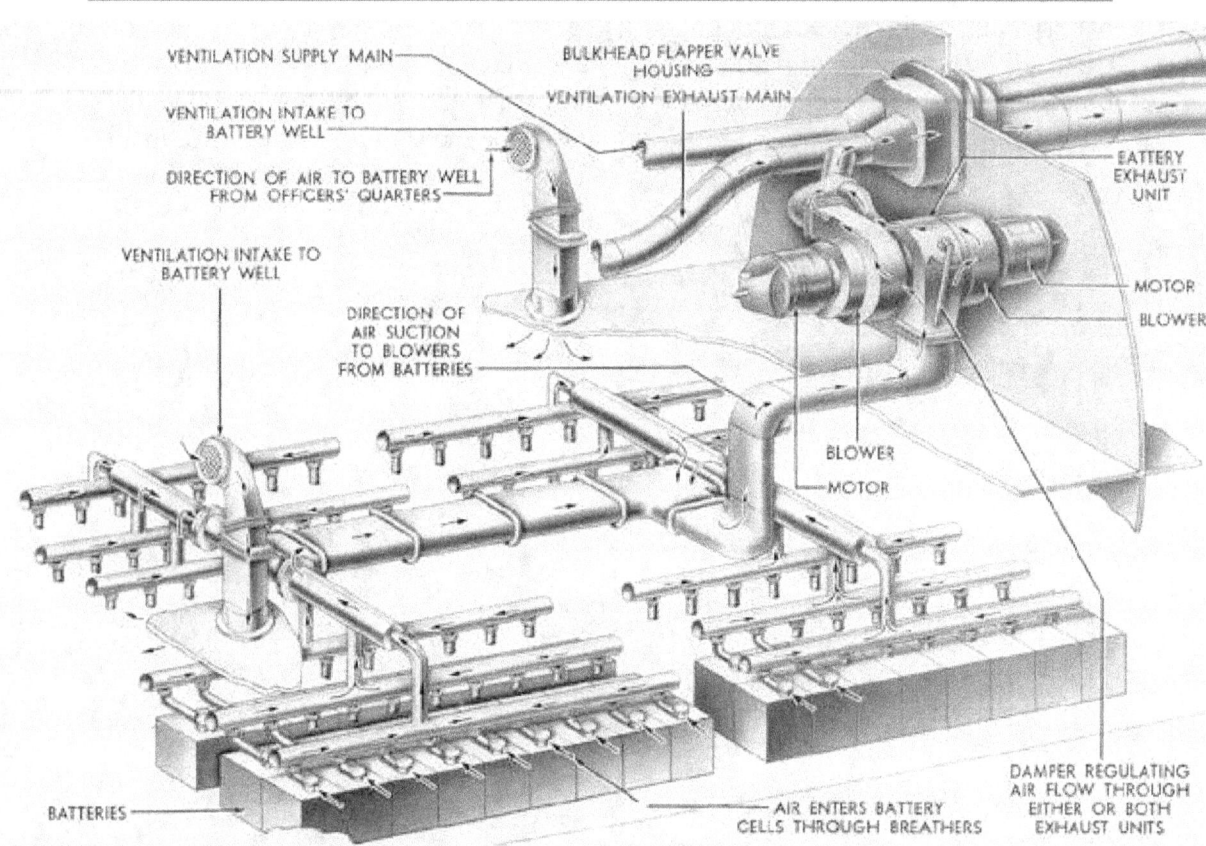

Figure 7-3. Battery ventilation.

operation are: a) when submerged, b) when using the air-conditioning coolers, and c) in case of damage to the outboard piping. When recirculation is taking place, all hull ventilation valves are shut. The ship's exhaust blower is discharging into the inlet of the ship's supply blower and the exhaust air from the after compartments of the ship is being picked up by the ship's supply blower. All this is accomplished by means of two damper-controlled louvers in a series of pipes and fittings between the discharge of the ship's exhaust and the inlet of the ship's supply blowers. To permit passage of exhaust air back to bulkhead. Each pair of valves is operated by the lever on either side of the bulkhead. Exhaust air may travel forward from the after to the forward machinery compartments only by means of the watertight door which must be left open for this purpose.

The ship's ventilation supply and engine induction valve, located in the after end of the conning tower fairwater, is operated hydraulically or by hand and is locked in either the open or shut positions by hand operation. The hand gear consists of a hand crank with two handles which operate a worm and worm gear so arranged as to raise and

the forward machinery compartment, all bulkheads between the forward end of the after torpedo room and the machinery compartments must be open to the compartments surrounding them by means of open exhaust bulkhead valves or doors.

Two air-conditioning coolers for cooling and drying the air are installed in the supply lines; the larger forward one is located in the after end of the crew's quarters and the smaller after one is located in the forward end of the after machinery compartment. When the humidity in the vessel becomes excessive, the quantity of air passing through the cooler should be reduced in order that the temperature of the air that passes through the cooler may be lowered below the dewpoint and thereby increase the quantity of water extracted from the air. The coolers are provided with drains to collecting tanks or engine room bilges.

The supply and exhaust mains at the watertight bulkheads are attached to pressure-proof hoods surrounding lever-operated bulkhead valves on each side of the bulkhead. Each pair of valves is operated by the lever on either side of the bulkhead. No ship's exhaust main exists aft of the ship's exhaust blower, but provisions for allowing exhaust air in the after end of the ship to get back to the machinery compartments have been made. Each of the two watertight bulkheads at the two ends of the maneuvering room has a pair of light hoods surrounding lever-operated bulkhead valves on each side of the

lower the valve stem through the hull by a bell crank and slotted lever arrangement. A double-acting piston type of hydraulic gear is in the power position. The hand gear also moves the hydraulic piston and can be used only when the control lever for engine induction on the flood control manifold is in the neutral position.

External gagging on the engine induction and ship's supply outboard valve is accomplished by a wrench-operated valve stem set flush with the deck. The valve stem is supported by a yoke superimposed on the valve body and is protected by a cover projecting slightly above the deck. Gagging of the engine induction and ship's supply outboard valve gags or *position locks* all internal operating gear. The operating gear for the engine induction and ship's supply outboard valve is fitted with a contact maker for the indicator lights in the control room, thus indicating the position (open or shut) of the valve.

There are four hull valves: one ship's supply, two engine induction, and one maneuvering room (auxiliary engine) induction. All four valves are of the flapper type and are gagged from the inside of the ship. All hull valves seat with pressure in the external piping

The operating gear for each of the following hull valves, one ship's supply, two engine induction, consists of an operating lever and a quick releasing gear located at a suitable distance from the valve and connected to it by means of intermediate levers and connecting rods. For opening each valve, the operating lever must be used, and may be used for closing to ease the valve to its seat after tripping. However, the valve can be shut in an emergency by squeezing the handles, thus tripping the quick-releasing mechanism which permits the valve to seat by its own weight.

The operating gear for each hull valve is fitted with a contact maker for indicator lights in the control room and engine rooms to show open and shut positions.

7A7. Hydrogen detecting systems. There are two types of hydrogen detectors in service; one is manufactured by the Cities Service Company (type N.H.D.) And the other by the Mine Safety Appliance Company (type M.S.A.). The function of the detectors is to take a sample of exhaust air continuously from the batteries and indicate the percentage of hydrogen concentration in the battery ventilation ducts.

The operation of both types of detectors is based on the principle of a balance Wheatstone bridge circuit. The air sample is drawn, by means of a motor-driven pump, across one leg of the balanced circuit where it is caused

to burn with an intensity dependent upon the amount of hydrogen present. The heat created heats the leg and increases its resistance, thereby creating an electrical unbalance in the entire circuit. The meter connected across the bridge circuit then shows a deflection on a properly divided scale which is directly proportional to the percentage of hydrogen present in the air sample.

In addition to the meter indication, the M.S.A. type has a white light connected in the circuit, indicating normal operation as long as the hydrogen content is below 3 percent. When the meter pointer indicates 3 percent on the scale, the circuit to a red warning light is closed. This red warning light will remain ON despite a decrease in hydrogen content until manually reset. Both meter and light indications are transmitted to repeater instruments in the maneuvering room.

The type N.H.D. detector is supplied with 115- to 120-volt alternating current directly from the a.c. bus of the I.C. switchboard. This system uses a rectifier to convert the alternating current into direct current for the bridge circuit.

The M.S.A. type detector is supplied with 120-volt direct current from the lighting feeder.

B. AIR PURIFICATION

7B1. General description. Air purification is accomplished by the use reaching 3 percent at the time of surfacing.

of a CO_2 absorbent as outlined in the Bureau of Ships instructions. Thirty-seven canisters, containing 15 pounds each of CO_2 absorbent, are carried in stowages distributed in the several living compartments.

The limiting percentage of CO_2 is 3 percent. One percent or less is harmless, and after air purification is started, efforts should be made to keep the percentage of carbon dioxide from going above this amount. If, in any case, it becomes necessary to conserve the CO_2 absorbent, the percentages of carbon dioxide may be allowed to increase during the last few hours of submergence, barely

Two percent of carbon dioxide will ordinarily not be noticed, but may show some discomfort if work requiring strenuous exertion is attempted. Prolonged breathing of over 3 percent CO_2 causes discomfort in breathing even at rest and becomes progressively dangerous above 4 percent. The amount of carbon dioxide should never be allowed to exceed 3 percent. If, for any reason, it does reach this concentration, it should be reduced as rapidly as possible. The limiting percent of oxygen (O), on the other hand, should not fall below 17 percent.

To maintain the air of a submarine within these limits of purity, it is necessary to

reduce the increasing CO_2 by chemical absorbents and to replenish the decreasing supply of oxygen by bleeding certain amounts of oxygen or air into the submarine at regular intervals.

After the original enclosed air in the submarine has become so vitiated that the limiting values of oxygen and carbon dioxide have been reached, one of the following procedures is necessary to revitalize the atmosphere:

a) Bleed into the vessel 0.9 cubic foot of oxygen at atmospheric pressure per man per hour and at the same time use the Navy standard carbon dioxide absorbent.

b) In lieu of oxygen, bleed into the vessel air from the compressed air tanks at the rate of 31 cubic feet of air

c) Start the high-pressure air compressors and pump a slight vacuum in the submarine, charging this air into a low air bank; then release air into the submarine from a bank that was charged on the surface.

Before submerging, whenever circumstances permit, thoroughly ventilate the vessel by closing the hatches and ventilators in such a manner that all the air for the engines is drawn into the end compartments and through the vessel. Under these conditions, run the engines for 5 minutes.

Since the contained air in the vessel at the outset of the dive is thus pure, it will not reach the limiting values of 17 percent O and 3 percent CO_2 until the expiration of a period of hours (X), calculated by the following formula:

at atmospheric pressure per man per hour. The introduction of additional air or oxygen is solely to replenish the oxygen content in the compartment, which, under normal submerged operating conditions, should not be allowed to fall below 17 percent. Irrespective of whether compressed air or compressed oxygen is used for this purpose, the CO_2 absorbent should also be used, in view of the fact that exhaustive tests have indicated that the deleterious physiological effects of high concentrations of CO_2 are not alleviated appreciably by the introduction of additional oxygen or air.

$$X = 0.04 \text{ C/N}$$

where C = net air space of the submarine in cubic feet, and N = the number of men in the crew.

If submergence under ordinary operation conditions is less than 17 hours, oxygen or compressed air replenishment and CO_2 purification should not be necessary. However, if it is predetermined that the time of submergence will be greater in any case than 17 hours, air purification with the CO_2 absorbent should be resorted to at the end of the periods indicated for the respective class of ship.

CO_2 absorbent is considerably more expensive than soda lime and great care should be exercised in the handling and stowage of the containers. Exposure of them to external pressures, such as are employed in testing compartments, would probably rupture their seams and destroy their airtightness, thus causing eventual deterioration of the absorbent. The containers should be removed during air testing of compartments.

Figure 7-4. Carbon dioxide absorbent.

This chemical absorbs CO_2, by contact. The larger the area of the exposed surface of the absorbent, the more efficient will be the result. When the length of submergence is such as to necessitate CO_2 elimination, the following steps should be taken:

a. Remove the mattress covers from the

mattresses of four lower bunks in the most convenient compartment provided with outboard ventilation when surfaced.

amount of heat evolved will be slight. Consequently, when determining the warmth of the chemical by touch, care should be taken that the material is fully

b. Slit the mattress covers and spread the covers, single thickness, as smoothly and taut as possible over bunk springs. Lash the edges to the bunk spring, if necessary, to keep it taut. Remove the cover from one of the CO_2 absorbent containers and pour about one-fourth of the contents (approximately 3 1/2 pounds) on the cover. With a stick, spread the chemical as evenly as possible over the mattress covers. In pouring the chemical from the container and in spreading it on the mattress covers, care should be taken not to agitate it any more than necessary, as it is caustic and the dust will cause throat irritation. The irritation, however, is only temporary and while in many instances, coughing and sneezing may be induced, the effects are not harmful. After working with the chemical, do not rub the eyes before the hands have been thoroughly washed. If it should get into the eyes, painful, but not dangerous, irrigation will result.

It may be relieved by washing the eyes with a solution of 1 part vinegar or lemon juice and 6 parts of water, or by careful washing with a quantity of fresh water. Do not spread the chemical with the hands. Use a stick or other means. After spreading, stir it gently once each hour.

Under normal submerged operation conditions, the contents of one container when spread on four mattress covers, as outline above, will absorb CO_2 for 144 man hours, or will absorb the CO_2 produced by a crew of 33 men for approximately 4 1/2 hours; a crew of 43 men for approximately 3 1/2 hours; a crew of 87 men

spent before renewing it. If there is any doubt on this point, leave the material spread on the mattress cover and spread an additional charge on a split mattress cover in an additional bunk.

Because of the desirability of keeping the chemical from *dusting* as much as possible, the refilling of the mattress cover and the spreading of the chemical should not be resorted to any oftener than necessary. Accordingly, if submergence of any particular vessel is to exceed the time of CO_2 protection afforded by the chemical in the initial container, the contents of the necessary additional container or containers should be spread out on additional mattress covers in the amount necessary to furnish the increased man hours of protection required, in the same or other convenient compartments, using approximately 3 1/2 pounds of the chemical on each additional cover. The number of additional covers so used should depend upon the total number of men on board and the estimated total time for which protection from CO_2 must be afforded.

Higgins and Marriott carbon dioxide testing outfits are supplied for determining the percentage of carbon dioxide in the air on submarines, and one of the outfits should be carried on each submarine. The test with this apparatus is extremely simple and sufficiently accurate for all practical purposes.

7B2. Oxygen system. Eleven oxygen standard containers are distributed throughout the compartments of the vessel, the total capacity being sufficient to supply 37 cubic feet per man of oxygen at normal temperature and approximately 1 3/4 hours.

When this chemical absorbs CO_2, it evolves heat and is warm to the touch. The amount of heat evolved depends upon the amount of CO_2 present in the air and the rate of its absorption. When the chemical no longer evolves heat in the presence of CO_2, it has become saturated and should be renewed. However, with small numbers of men in a compartment, the amount of CO_2 generated will not be so great as that produced by a large number of men and the

atmospheric pressure. Two flasks are stowed in the forward torpedo room and two in the after torpedo room. Each of the other compartments, including the conning tower, has one flask. The flasks and regulator valves in the forward torpedo room, after torpedo room, and the conning tower are piped to form banks in each of the three compartments with a valve for the compartment and a manifold for escape arrangements.

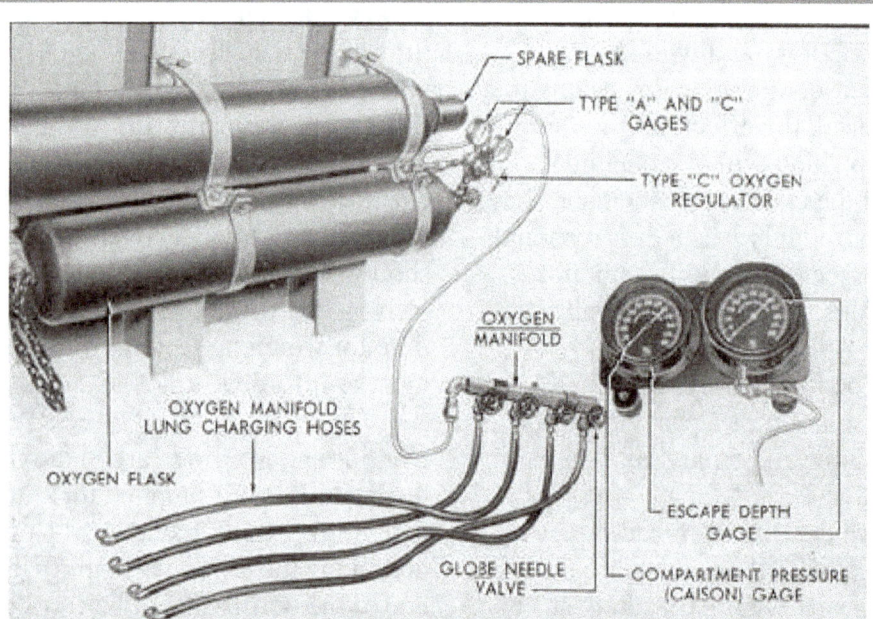

Figure 7-5. Oxygen tank.

These manifolds, located in the forward escape trunk, conning tower, and just forward of the escape trunk in the after torpedo room, have four valves. Each valve is provided with a 6-foot length of rubber hose, fitting with a self-closing chuck of the Schrader type, for charging escape lungs.

The containers in the other rooms are fitted with the regulator valves for replacing the oxygen content of the air as desired.

The container or containers in one compartment are not connected with those in another compartment.

C. VALVES

7C1. Classifications. The ventilation system is provided with valves classified as follows:

1. Engine induction and ship's supply outboard valve. (Superstructure abaft conning tower.)

2. Engine induction hull valve. (Forward engine room.)

3. Engine induction hull valve. (After engine room.)

4. Maneuvering room induction hull valve. (Maneuvering room.)

5. Ship's supply hull valve. (Forward engine room.)

6. Bulkhead flapper valves, supply and exhaust.

All of these valves are described separately in the following sections.

7C2. Engine induction and ship's supply outboard valve. The engine induction and ship's supply outboard valve is a 36-inch disc type valve (Figure 7-1), located in the air induction standpipe in the superstructure abaft the conning tower. (See Figure A-10.)

When open, this valve permits air to enter the engine air induction and the hull ventilation supply lines; when shut, it

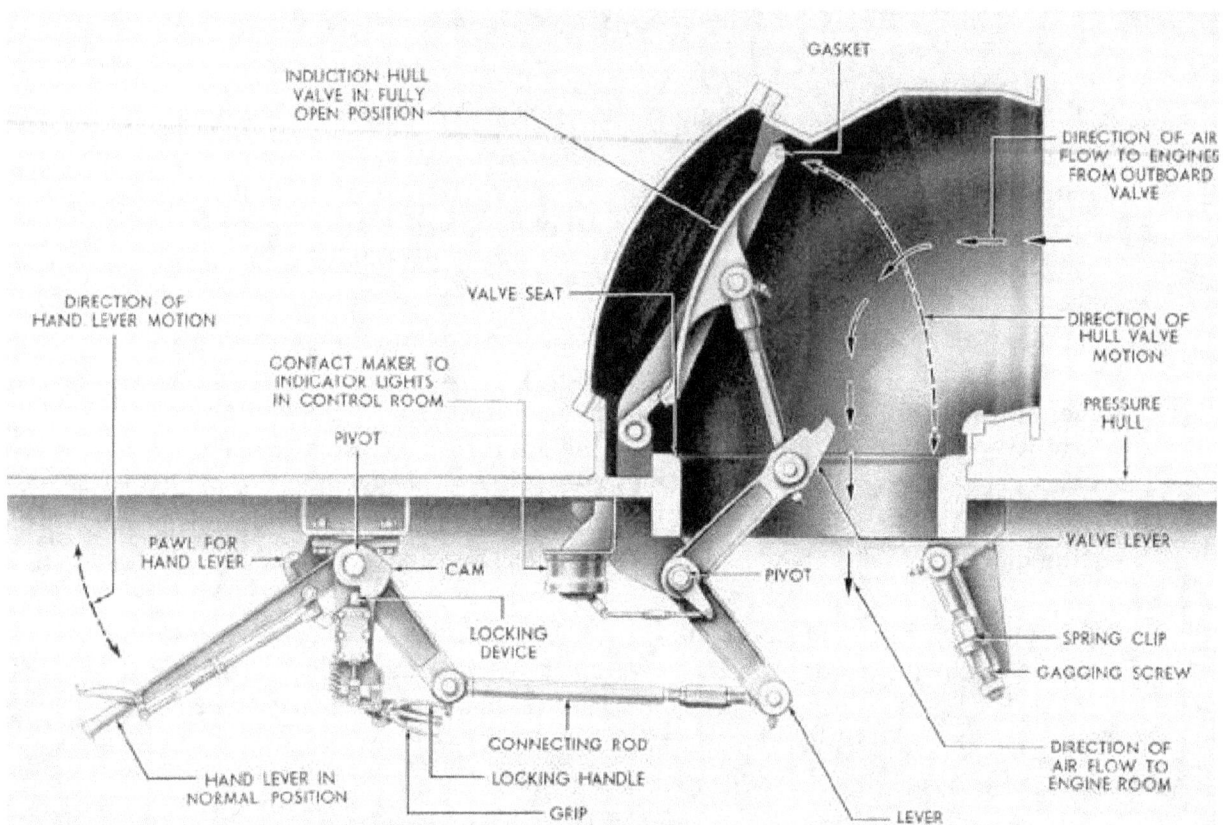

INDUCTION HULL
VALVE IN FULLY
OPEN POSITION

GASKET

DIRECTION OF AIR
FLOW TO ENGINES
FROM OUTBOARD
VALVE

VALVE SEAT

DIRECTION OF
HAND LEVER MOTION

DIRECTION OF
HULL VALVE
MOTION

CONTACT MAKER TO
INDICATOR LIGHTS
IN CONTROL ROOM

PRESSURE
HULL

PIVOT

PAWL FOR
HAND LEVER

CAM

PIVOT

VALVE LEVER

LOCKING
DEVICE

SPRING CLIP

GAGGING SCREW

CONNECTING ROD

DIRECTION OF
AIR FLOW TO
ENGINE ROOM

HAND LEVER IN
NORMAL POSITION

LOCKING HANDLE

GRIP

LEVER

Figure 7-6 Engine induction hull valve.

86

prevents the entrance of air or water to all three lines. It is operated hydraulically from the control room and is open only when the vessel is on the surface. It can be operated *manually* from the crew's mess.

7C3. Engine induction hull valve. There are two main engine induction hull valves located beneath the superstructure at those points where:

a. The port engine air induction line goes through the pressure hull into the forward engine room.

b. The starboard engine air induction line goes through the pressure hull into

valve obtains its supply of air from a damper fitting in the starboard engine induction line located in the superstructure. The maneuvering room induction hull valve receives bypassed air to cool the operating control stand in the maneuvering rooms. The amount of air that is bypassed is controlled from the after engine room by a lever-operated damper in the damper fitting.

7C5. Ship's supply hull valve. The ship's supply hull valve obtains its supply of air from the engine induction and the ship's supply outboard valve. The valve is similar in construction to the engine induction hull valve shown in the after engine room.

These valves are used to direct the flow of air from the engine air induction and ship's supply outboard valve to the forward engine and after engine rooms. Each is manually operated from within the pressure hull.

Figure 7-6 shows the general construction of a typical engine induction hull valve.

Each valve is of the quick-closing type, set into an angle housing. The outboard, or inlet side of the housing connects to the supply lines, bringing air from the engine air induction and ship's supply outboard valve in the superstructure. The inboard, of discharge side, leads into the pressure hull.

The valve is operated by a linkage extending to a hand lever. To open the valve, the handle is moved downward until the cam locking device is set. This will hold the valve in the open position. With the cam locking device set, the valve handle is returned to its former position. The valve, however, is held open by a quick-releasing cam locking device on an *open hair trigger* for instant release and quick closing when the locking handle is depressed. This withdraws the locking device, releases the cam, and allows the valve to seat by its own weight. A link on the moving arm actuates a switch which, in turn, operates a signal light in the control room, indicating the position of the valve.

7C4. Maneuvering room induction hull valve. The maneuvering room induction valve is similar to but

Figure 7-6.

7C6. Bulkhead flapper valves. At each point where the hull supply and hull exhaust lines pass through a bulkhead, two bulkhead flapper valves are installed. There is no exhaust line aft of the forward engine room and therefore no exhaust flapper valve between the forward engine room and the after engine room bulkhead.

Figure 7-7 shows in schematic form the general construction and arrangement of a typical set of bulkhead flapper valves.

One valve is mounted on either side of the bulkhead, with the pivot shaft projecting downward and fitted with a handle. The two pivot shafts are interconnected by gears so that moving one shaft and valve will operate the other simultaneously. This arrangement permits operating the valves from either side of the bulkhead.

A locking device, extending through the bulkhead, permits locking or unlocking of the operating handles from either side of the bulkhead.

When shut, the bulkhead flapper valve stops all ventilating air flow to the compartments forward or aft, as the case may be, of the bulkhead on which the shut valve is mounted.

Bulkhead flapper valves are also provided on all bulkheads aft of the forward engine room except the bulkhead between the forward and after engine room. These valves permit free exhaust air to flow to the engine rooms. All bulkhead flappers seat with compartment pressure.

smaller than the engine induction hull
valve shown in Figure 7-6. This

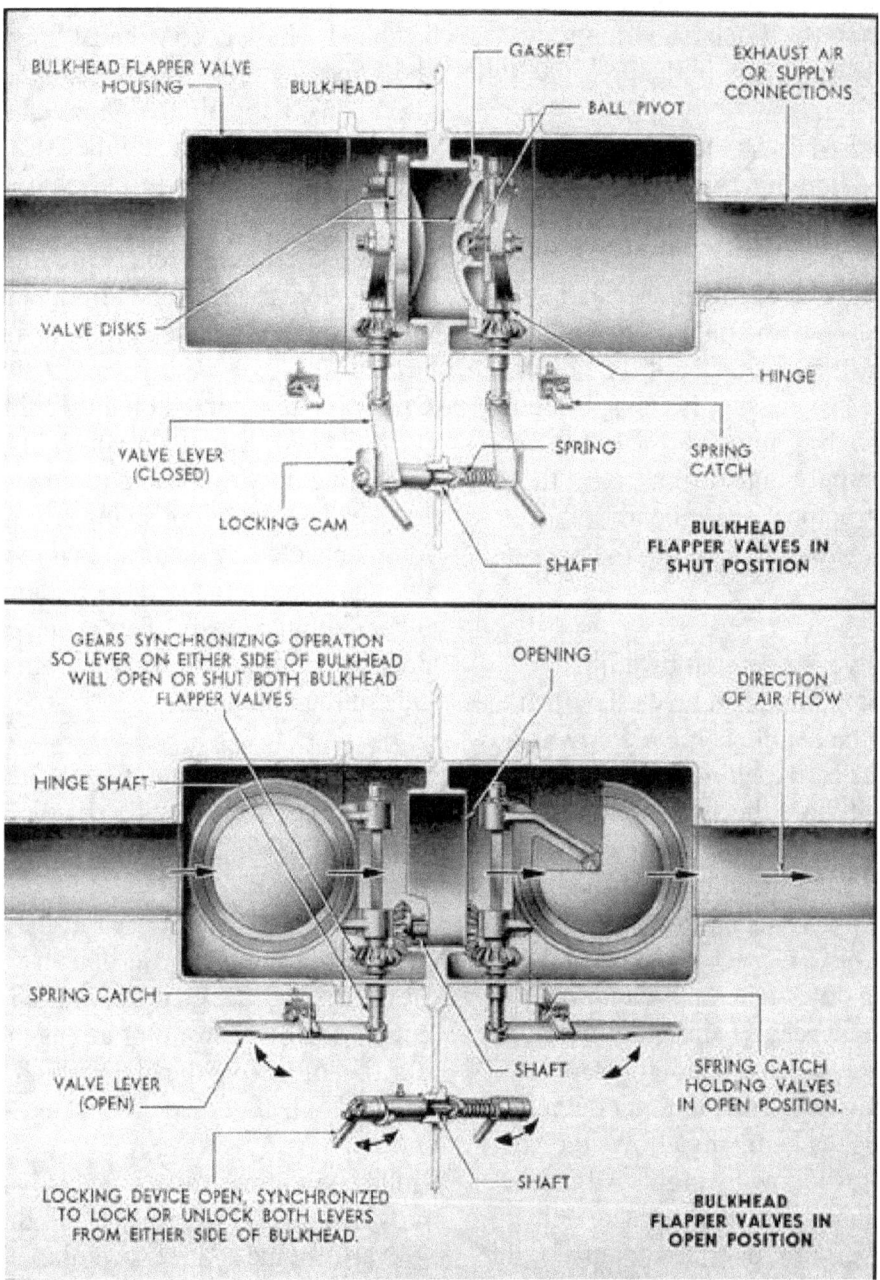

Figure 7-7. Bulkhead flapper valves.

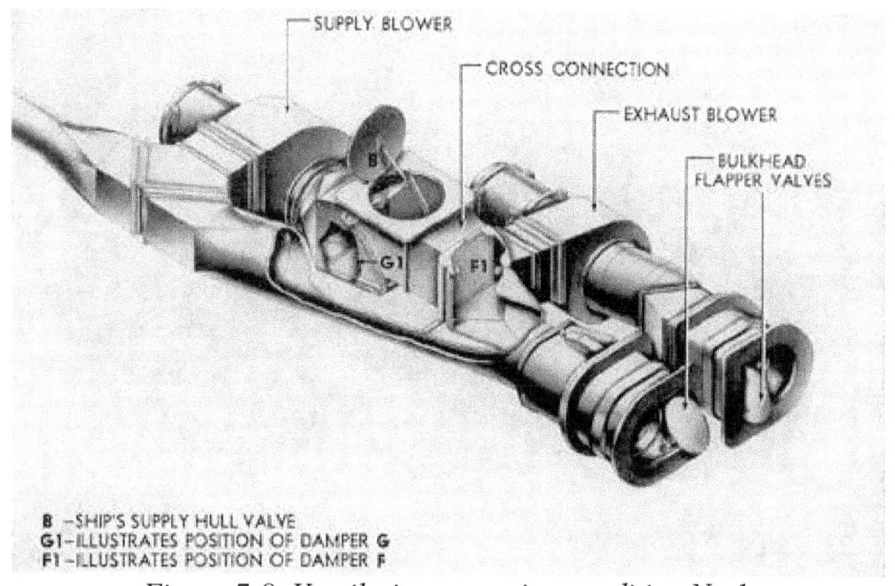

B —SHIP'S SUPPLY HULL VALVE
G1—ILLUSTRATES POSITION OF DAMPER G
F1—ILLUSTRATES POSITION OF DAMPER F

Figure 7-8. Ventilation operation condition No 1.

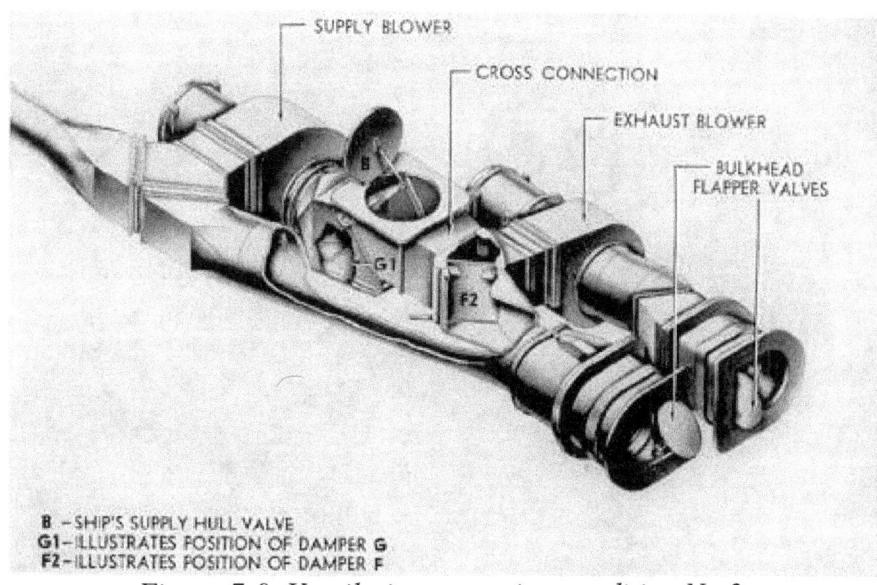

B —SHIP'S SUPPLY HULL VALVE
G1—ILLUSTRATES POSITION OF DAMPER G
F2—ILLUSTRATES POSITION OF DAMPER F

Figure 7-9. Ventilation operation condition No 2.

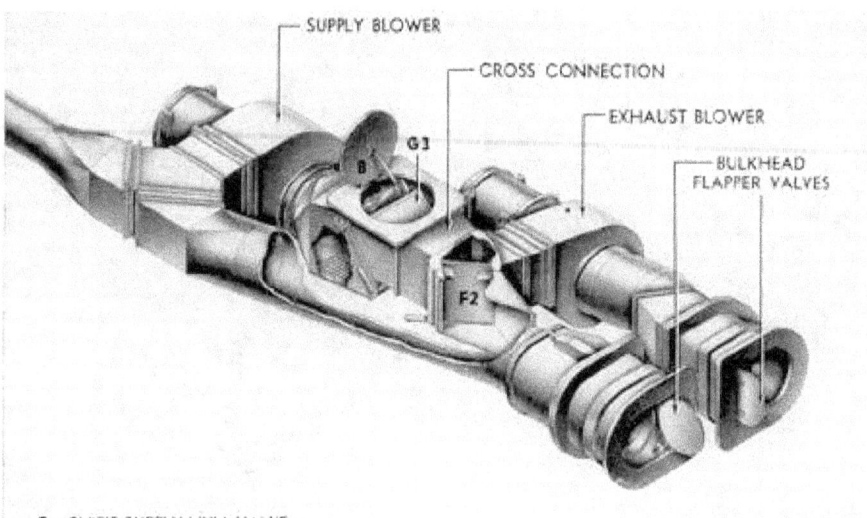

SUPPLY BLOWER
CROSS CONNECTION
EXHAUST BLOWER
BULKHEAD FLAPPER VALVES
G1
B
F2

B – SHIP'S SUPPLY HULL VALVE
G3–ILLUSTRATES POSITION OF DAMPER **G**
F2–ILLUSTRATES POSITION OF DAMPER **F**

Figure 7-10. Ventilation operation condition No 3.

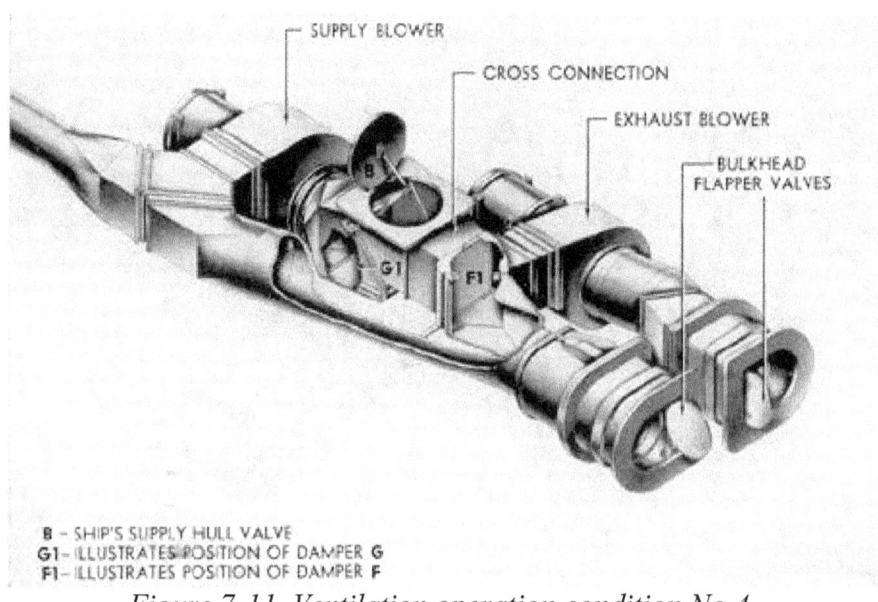

SUPPLY BLOWER
CROSS CONNECTION
EXHAUST BLOWER
BULKHEAD FLAPPER VALVES
B
G1
F1

B – SHIP'S SUPPLY HULL VALVE
G1– ILLUSTRATES POSITION OF DAMPER **G**
F1– ILLUSTRATES POSITION OF DAMPER **F**

Figure 7-11. Ventilation operation condition No 4.

90

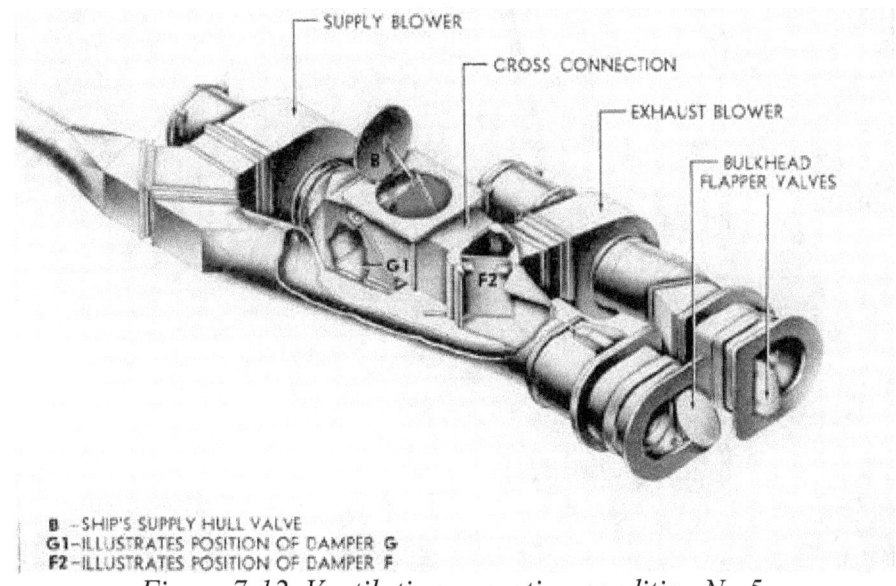

B —SHIP'S SUPPLY HULL VALVE
G1—ILLUSTRATES POSITION OF DAMPER G
F2—ILLUSTRATES POSITION OF DAMPER F

Figure 7-12. Ventilation operation condition No 5.

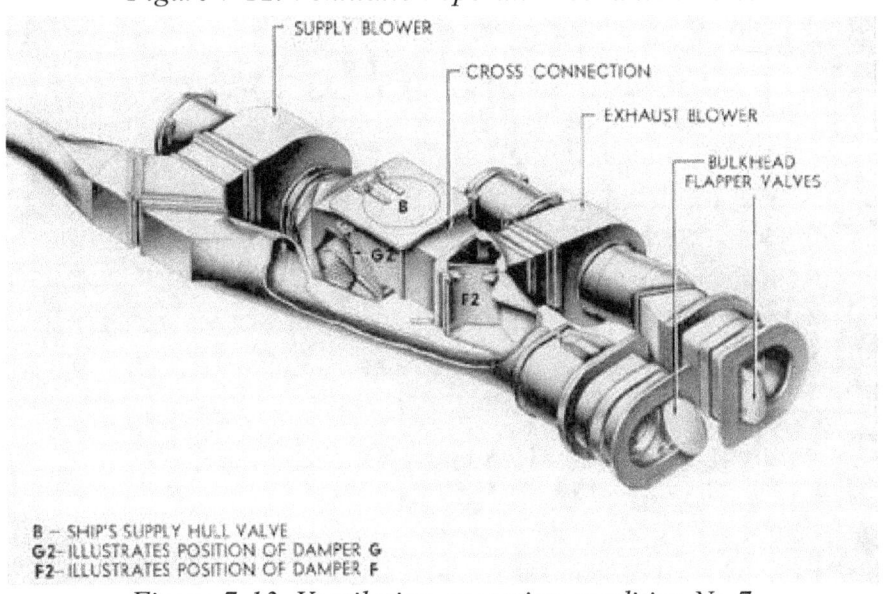

B — SHIP'S SUPPLY HULL VALVE
G2— ILLUSTRATES POSITION OF DAMPER G
F2— ILLUSTRATES POSITION OF DAMPER F

Figure 7-13. Ventilation operation condition No 7.

91

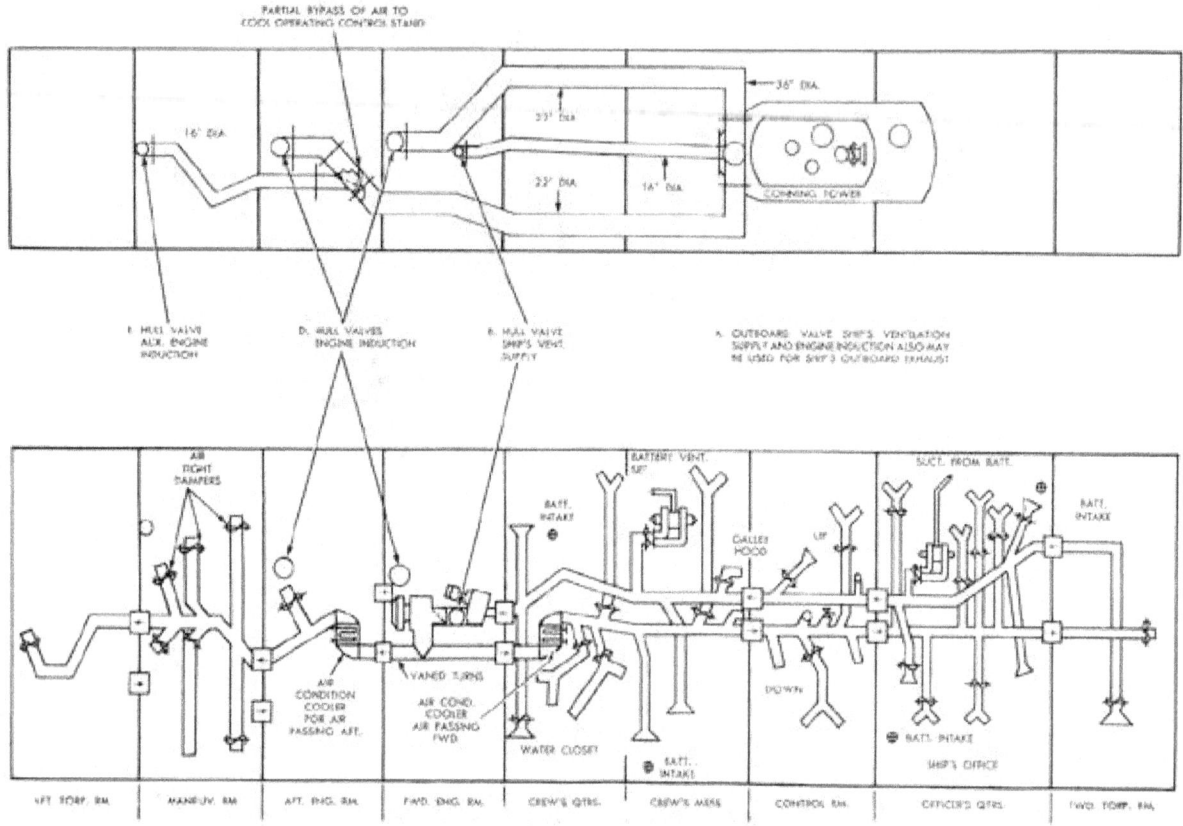

Figure 7-14. Schematic diagram of the ventilation system.

92

VENTILATION SYSTEM TABLE OF OPERATING CONDITIONS									
Number of Operation	AIR CONDITION	Position of valve and dampers						Condition of vessel	Condition of ventilation
		A	B	C	D	F	G		
1. Figure 7-8	Stop	Open	Open	Shut	Shut	F1	G1	On surface, engines STOPPED	Supply from outboard. Exhaust outboard via engine rooms and hatches.
2. Figure 7-9	Run	Open	Open	Shut	Shut	F2	G1	On surface, engines STOPPED	Recirculating and makeup air is from outboard.

3. Figure 7-10	Stop	Open	Open	Shut	Shut	F2	G3	On surface, engines STOPPED	Supply from inboard via hatches and engine rooms. Exhaust outboard via normal supply pipe.
4. Figure 7-11	Stop	Open	Open	Open	Open	F1	G1	On surface, engines RUNNING	Supply from outboard. Exhaust to engine room.
5. Figure 7-12	Run	Open	Open	Open	Open	F2	G1	On surface, engines RUNNING	Recirculating and makeup air is from outboard.
6.	Stop	Shut	Shut	Shut	Shut	F2	G2	SUBMERGED	Recirculating without air-conditioning.
7. Figure 7-13	Run	Shut	Shut	Shut	Shut	F2	G2	SUBMERGED	Recirculating with air-conditioning.

7C7. Chart of operating conditions. Valve and damper line-up for various operating conditions is shown on the chart on the preceding page. In interpreting this chart it is necessary to know the meaning of the following symbols which are used frequently:

KEY FOR TABLE OF OPERATING CONDITIONS	
Symbol	**Meaning**
A	Engine induction and ship's supply outboard valve.
B	Ship's supply hull valve.
D	Forward and after engine room, engine induction hull valves.
E	Maneuvering room induction hull valve.
F	Exhaust blower discharge damper.
G	Supply blower suction damper.
POSITION OF DAMPERS	

F1	Exhaust blower to discharge into engine room and cross connection shut off.
F2	Exhaust blower to discharge into cross connection and engine room shut off.
G1	Supply blower suction on cross-connection and engine room shut off.
G2	Supply blower suction on cross connection and partial suction on engine room.
G3	Supply blower suction on engine room and cross connection shut off.

8
REFRIGERATING AND
AIR-CONDITIONING SYSTEMS

A. PRINCIPLES OF MECHANICAL REFRIGERATION

8A1. Brief statement of principles.
The mechanical refrigeration method used on board a submarine is the vapor refrigerating process. In this process, the refrigeration passes alternately through its liquid and vapor states. Such a refrigerant, therefore, must have special qualities. It must boil at a very low temperature and it must be able to change its state from liquid to vapor and vice versa readily. Above all it must be a *safe* refrigerant. This is more important in submarines than in other types of vessels.

In the liquid state, the refrigerant picks up heat from substances or the air in spaces and in so doing, vaporizes. The vapor, carrying the excess head, is then moved away to another location where it gives up or discharges that heat, and is converted back to the liquid state.

The mechanical system in which the refrigerant is contained is a single airtight circuit of pipes and mechanisms through which the refrigerant is pumped continuously, so that a given quantity is used over and over. This requires an input of energy which is

from the point at which the heat to be removed enters the refrigerating system. This point is where the evaporator is located. Figure A-11 shows a simplified diagram of the main mechanical elements in the cycle.

8A3. Through the evaporator. The evaporator is simply a bank, or coil, of copper tubing. It is filled with Freon 12 at low pressure and temperature. Heat flowing from the air spaces or articles to be cooled into the coil will cause the liquid Freon to boil. Boiling can take place only by the entrance into the liquid of its latent heat of vaporization, and then this latent heat can come only from the surrounding substances. Hence, their temperatures are lowered. The latter portion of the evaporator coil is therefore filled with Freon vapor at low pressure, carrying with it the unwanted heat.

8A4. Through the compressor. This vapor does not remain in the evaporator. The compressor is operating and the suction that it exerts (on the evaporator side of its circuit) pulls the heat-laden vapor out of the evaporator, through the piping, and into the compressor. The

supplied by an electric motor. Figure A-11 shows the cyclical arrangement of the essential elements of the refrigeration system; namely, the evaporator, compressor, condenser, receiver, and thermostatic expansion valve. The liquid refrigerant picks up heat and vaporizes in the evaporator. The vapor then goes to the compressor where it is compressed to a pressure at which its temperature is above that of the water flowing through the condenser. The compressed vapor then passes to the condenser where sufficient heat is transferred to the water to cause the refrigerant vapor to condense. The condensed refrigerant, now a liquid, flows next to the receiver, and then through the thermostatic expansion valve to the evaporator.

8A2. The Freon 12 cycle. Let us follow through the cycle of operations, starting

compressor, therefore, is the mechanism that keeps the Freon in circulation through the system. In the compressor cylinders, the Freon is compressed from a low-pressure vapor to a high-pressure vapor and its temperature therefore rises.

8A5. Through the condenser. The Freon vapor, now at high pressure, passes next into the condenser, where the vapor passes around the tubes through which sea water is continuously pumped. Here the excess heat flows by conduction through the walls of the tubing from the higher-temperature vapor to the relatively lower-temperature sea water, and here, therefore, the unwanted heat leaves the primary refrigerating system and is finally carried away. This excess heat thus

flowing out of the vapor is latent heat of vaporization and therefore, the vapor condenses back to the liquid state. The liquid Freon is now at high pressure and high temperature.

8A6. Through the receiver. The liquid Freon passes next into the receiver, or tank. The liquid in this receiver acts as a seal between the vapor in the condenser and the liquid as it flows into the next element, the expansion valve, in order that the liquid Freon in the expansion valve may be free of vapor. The whole system is a single circuit in which the fluid is circulated.

8A7. Through the expansion valve. The liquid Freon enters the expansion

and low temperature, and is reentering the evaporator, its cycle completed, and ready to be repeated. Every part of the cycle is, of course, taking place simultaneously throughout the circuit, and continuously as long as the refrigeration is wanted. The entire operation is automatic.

8A8, The low-pressure side. That portion of the cycle from the orifice of the expansion valve around through the evaporator to and including the intake side of the compressor cylinders is called the low-pressure side. The dividing line between the low-pressure and the high-pressure sides is the discharge valve of the compressor.

valve at high pressure and high temperature. This valve regulates the flow of the refrigerant into the evaporator. The liquid outlet from the expansion valve is a small opening, or *orifice*. In passing through the orifice, the liquid is subjected to a throttling action, and is dispersed into a finely divided form. The Freon now is again a vapor at low pressure

8A9. The high-pressure side. The remainder of the cycle, that section from the discharge valve of the compressor around through the condenser, receiver, and expansion valve to its orifice is called the high-pressure side. The dividing line between the high-pressure and low-pressure sides is the thermostatic expansion valve.

B. MECHANICAL DETAILS OF AIR-CONDITIONING SYSTEM

8B1. The air-conditioning cycle. The air-conditioning cycle in the air-conditioning system is the same as in the refrigeration system. In general, the mechanical circuit of equipment is also similar, the main difference being that the air is brought by forced ventilation through ducts to the evaporators and returned through ducts to the rooms.

8B2. The air-conditioning plant. The air-conditioning plant consists of the following main elements:

a. Two compressors, York-Navy Freon 12, enclosed single-acting vertical, two cylinders, 4-inch bore x 4-inch stroke, rated at 4 refrigeration tons each.

b. Two condensers, York-Navy Freon 12, horizontal shell-and-tube 4-pass type.

c. Two receivers, York-Navy Freon 12 type.

d. Four evaporators, with finned cooling coils in two casings.

e. Two conning tower evaporators, in one casing.

8B3. Double system arrangement.

elements listed in Section 8B2 are connected as two separate systems, each containing all necessary valves, gages, and controls for automatic operation. The cooling coils of these two systems, however, are placed side by side in an evaporator casing, and though appearing to be a single unit of coils, are, nevertheless, entirely separate. Thus, either of the two systems may be operated alone, with its cooling action taking place in the evaporating casing.

8B4. Interconnection of double system. The double systems, while ordinarily set to operate individually, are interconnected. On the 200 class submarines, the interconnecting pipes run between a) the discharge lines of the condensers, b) the outlet lines of the condensers, and c) the inlet or suction lines to the compressors. Shutoff valves in these interconnecting pipes permit any of the main elements to be cut out of one system and put into the other, in case of necessity. There would be no flow in the interconnecting pipes unless their shutoff valves were opened; normally they are closed. On the 300 class

The main

submarines, the interconnecting pipes run between a) the discharge lines of the compressors and b) the outlet lines of the condensers. There is no interconnection between the suction lines of the compressors.

8B5. Capacity of the air-conditioning system. The capacity of the air-conditioning system is 8.0 tons of refrigeration with the two compressors operating at 330 rpm.

8B6. Necessity for compressors of different capacities. The air conditioning system and the refrigeration system aboard a submarine are designed as two separate and distinct systems. Each is capable of performing its task independently of the other. However, in practice it is desirable that these two systems be interconnected so that the air-conditioning compressor can serve as a standby for the refrigeration system. This will insure continuous operation of the refrigerating system in the event of a prolonged repair job on the refrigerating compressor, which otherwise would result in the spoilage of the food stored in the refrigerating rooms.

In earlier references and explanations within this text, the rated capacity of the refrigeration system compressor has been given as 1/2 refrigeration ton, while the rated capacity of each of the air-conditioning system compressors is given as 4 refrigeration tons. This difference in rated capacity of the two units is due to the fact that the air-

These requirements determine the work load of each system, and this work load, expressed in refrigeration tons, determines in turn the capacity of the compressor needed for each system.

8B7. Relation of capacities. The capacity of the compressor is the amount of work, expressed in refrigeration tons, that a compressor is capable of performing under a single set of operating conditions. A change in the operating conditions will cause a corresponding change in the rated capacity of the compressor. Therefore, the relation between the capacity of a compressor on a refrigeration system and one on an air-conditioning system is a comparison of the operating conditions of evaporator temperature, speed of the compressor, and temperature of the cooling medium for each system, and not a comparison of the compressor or its maximum work load under optimum conditions.

The misunderstanding of this relationship has often given rise to a question as to whether or not there is difference between a *ton of air-conditioning* and a *ton of refrigeration.* The cause of this question is the apparent increase in the capacity in refrigeration tons developed by a compressor on an air-conditioning unit over the capacity developed by the same compressor when operating on a refrigerating unit. Although it may appear that there is a difference between a *refrigeration ton* and an *air-conditioning ton,* actually there is none and the term *air-conditioning ton* is not

conditioning system performs a greater amount of work than does the refrigeration system. The refrigeration system has to perform only sufficient work to remove heat from the comparatively small space of the cool and refrigerating rooms, with the minor addition of the ice cuber. Since these two rooms are thoroughly insulated, little or no heat enters them from the outside. The only source of heat, therefore, inside the rooms is from the stowed foodstuffs and from persons entering for supplies. The air-conditioning system, on the other hand, has to remove heat generated throughout the ship. This is heat which passes into the air of the ship from the engines, crew, cooking, batteries, electric light bulbs, equipment, and at times from the surrounding water outside the hull.

in acceptable usage.

The basic rating of refrigeration ton or heat-removing capacity of a machine is exactly the same whether the machine is used for removing heat from an icebox or lowering the humidity and/or temperature of the air in a submarine. However, a compressor that is rated at 2.95 refrigeration tons when operating at a -5 degree Fahrenheit evaporator temperature, running at 600 rpm with the same temperature of cooling water in the condenser in both cases will have a rating of 8,348 refrigeration tons if it is operated at a 35 degree Fahrenheit evaporator temperature. Thus, the rating of a compressor may vary, depending upon the evaporator temperature; also, the rating of a compressor may vary, depending upon

the speed of the compressor and the temperature of the cooling water flowing through the condenser. The operating temperature of the evaporator (suction pressure) will have the greatest effect on the number of refrigeration tons that the compressor will develop. The higher the suction (and pressure), the smaller the pressure differential between the suction and discharge; hence the compressor will handle more Freon with less work. In other words, the compressor will handle approximately twice the gas at 40 degrees that it will at 0 degrees Fahrenheit. Therefore, more refrigeration tons are developed at the higher evaporation temperatures (suction pressure).

8B8. Cross connection of air-

pressure will be maintained in the air-conditioning evaporator while the compressor is operating on the lower suction pressure necessary on the refrigerating system. The operation of the two systems in this manner is desirable because the capacity of the air-conditioning compressor is much greater than is needed to maintain the refrigerating rooms at their desired temperatures.

The current supplied to the thermostatic control on the refrigeration system is another point that must be checked, otherwise the solenoid valves will remain closed and no refrigerant will flow through the system. On some ships, the thermostat circuits are energized from 110 volt d.c. so that in this case the main thermostatic control circuit will still be

conditioning and refrigeration system. In an emergency it is possible to cross connect at least one of the air-conditioning compressors, condenser, and receiver, to the refrigerating system evaporator and maintain the desired temperatures in the refrigeration rooms.

On some classes of submarine, either of the air-conditioning compressors may be cross connected to the refrigeration system; on other classes only the No. 1 air-conditioning compressor can be used. As the arrangement and location of valves and lines vary on each installation, no detailed description can be given here. It is never necessary nor desirable to cross-connect the refrigerating compressor to the air-conditioning evaporators.

When cross-connecting the air-conditioning compressor to the refrigerating evaporator, there are several major adjustments that must be made on all installations. The low-pressure cutout on the air-conditioning compressor must be reset so that it will not stop the compressor until the suction pressure drops down to 2 psi. Normally this cutout is set to stop the compressor when the suction pressure on the air-conditioning evaporator reaches 32 psi. If the compressor is to operate both the air-conditioning system and the refrigerating system, the bypass around the suction pressure regulating valve should be closed, the stops opened, and the valve cut into the system. With the suction regulating valve in operation, a 32-pound suction

energized. On some 300 class submarines, the thermostat circuits are energized through the refrigerating control panel and when the main switch is pulled on the refrigerating compressor, it interrupts the supply to the thermostatic control circuits. In this case, the following procedure should be followed:

Leave in the main switch supplying current to the refrigerating compressor; with some insulated material, lift the overload relay cutout located on the bottom left side of the half-ton compressor control panel, to the OFF position, making sure that the overload relay cutout stays up. Then turn the selector switch on the half-ton system to either MANUAL or AUTOMATIC. This will insure a supply of current to the thermostatic controls.

8B9. Cross connection of air-conditioning systems. On some classes of submarines, it is possible to operate the No. 1 compressor and condenser on the No. 2 evaporator and vice versa. The air-conditioning systems on the 300 class submarines are cross-connected only by the compressor discharge lines and the high-pressure liquid lines. There is no cross connection between the suction lines. Because of this arrangement in the air-conditioning system, the No. 1 compressor can be connected to the No. 2 condenser, and the No. 2 compressor can be connected to the No. 1 condenser. The No. 1 compressor cannot be connected to the No. 2 evaporator, nor the No. 2 compressor to the No. 1 evaporator.

WATER SYSTEM

A. INTRODUCTION

9A1. General. Although the, submarine operates in large bodies of sea water, the use of salt water aboard the submarine is limited. Water, free of salt and other impurities, is used for cooling the diesel engines and in the crew's cooking, drinking, and bathing facilities. The torpedoes and the torpedo firing mechanisms, as well as the vacuum pump tank, use distilled water. Distilled water is also needed for the battery cells.

The water for all these operations is either carried by the vessel or is distilled on board. The purpose of the water system is to store, distill, and distribute water to the equipment requiring it. (See Figure A-8.)

B. FRESH WATER SYSTEM

9B1. General description. Two of the four main tanks of the fresh water system are located in the forward end of the forward battery compartment, and two in the after end of the control room below the platform deck. Fresh water tank No. 1 is located between frames 35 and 36 on the starboard side, tank No. 2 between frames 35 and 36 on the port side, tank No. 3 between frames 57 and 58 starboard, and tank No. 4 between frames 57 and 58 port.

The fresh water tanks are connected by means of the fresh water filling and transfer lines. Supply branches connect to the three emergency tanks, lavatories, sinks, showers, scuttlebutts, galley equipment, distilling plant, and the diesel engines. A cross connection is provided between the filling and transfer lines of the fresh water system and the filling and transfer line of the battery system.

Two 60-gallon emergency fresh water tanks are located on the port side in the forward torpedo room. One 130-gallon emergency fresh water tank is located on the port side in the after torpedo room. These tanks are connected to the fresh water system, while the 18-gallon emergency fresh water tank in the control room, and the 8-gallon emergency fresh water tank in the maneuvering room have no connections to the fresh water systems.

The fresh water filling valve and hose connection (Figure 9-1), located in the gun access hatch, connects with the fresh water filling and transfer lines extending to the forward and after ends of the vessel. In the forward torpedo room, the fresh water main has connections to the No. 1 and No. 2 fresh water tanks. It also has connections to the two 60-gallon emergency fresh water tanks, the crew's

Figure 9-1. Ship's fresh water filling valve.

lavatory, and the torpedo filling connections. The 60-gallon emergency fresh water tanks are equipped with their own torpedo filling connections. The quantity of water in the No. 1 and No. 2 fresh water tanks is measured by try cocks located on the after bulkhead of the forward torpedo room. (See Figure 9-2.)

In the officers' quarters, the fresh water main supplies fresh water to the officers' pantry, the shower, the lavatories, and the hot water heater.

Figure 9-2. Try cocks.

In the control room, the fresh water connections are to the fresh water tanks No. 3 and No. 4 below decks, and the fresh water transfer cutout valve from No. 3 and No. 4 fresh water tanks. The cross-connection valve between the fresh water system and the battery fresh water system is also in the control room overhead.

In the crew's quarters, the fresh water

to the after engine cooling system and purifiers.

There are no fresh water connections in the maneuvering room. The fresh water main aft terminates in the after torpedo room where it supplies water to the emergency fresh water tank, the after torpedo filling connection, and crew's lavatory. The emergency fresh water tank is equipped with its own after torpedo

supply connections are to the galley equipment, scuttlebutt, scullery sink, and coffee urn. In the after end of the crew's quarters, the fresh water main supplies water to the two lavatories, showers, and the hot water heater.

The water main in the forward engine room is equipped with valves and connections to the distilling plants, and to the forward engine cooling system and purifiers.

In the after engine room the fresh water main is equipped with valves and connections

9B2. Hot water system. Water for washing and cooking is heated by electric heaters. There are three electric hot water heaters. One heater with a 20-gallon tank is located in the starboard after corner of the control room; two heaters each with 25-gallon tanks are located one in the starboard forward corner of the forward battery compartment, and the other in the port after corner of the after battery compartment. Each heating unit is supplied with cold water from the fresh water mains.

filling connection.

C. BATTERY WATER SYSTEM

9C1. Purpose. The cells of the forward and after storage batteries must be filled periodically to maintain a safe level of liquid. The time between fillings is dependent upon battery use and operating conditions. The water used in the battery cells must be free of minerals and impurities which, while harmless to human beings, may react with the battery acid and plates to cause corrosion and breakdown of the battery cells. Therefore, only the purest distilled water may be used for refilling the batteries.

The purpose of the battery water system is to store and supply distilled water to the forward and after batteries.

9C2. Description and operation. The battery water system consists essentially of two groups of four tanks each, filling and transfer lines, and valves and branch piping with hose

forward battery water tanks with No. 5 and No. 7 battery water tanks located on the starboard side, and No. 6 and No. 8 tanks on the port side. The battery water filling valve and hose connection are located in the gun access trunk. A cross connection in the control room enables the battery water system to be supplied from the fresh water system. The battery water filling line divides into the forward and after supply lines, supplying water directly to their respective tanks. The supply line to the after battery water tanks is connected to the distilling plant, providing an additional supply of distilled water for the batteries when the distilled water in the battery water tank is consumed.

The battery cells are filled by means of a hose which is attached to the battery filling connection, located on the battery filling line connecting the port and starboard tanks. The tanks are so

connections for filling the individual battery cells. The four forward battery water tanks, Nos. 1, 2, 3, and 4, are located below deck in the forward battery compartment and are arranged in tandem, two on the port side and two on the starboard. The after battery water tanks are arranged similarly to the interconnected that any one of the tanks can be used to supply the cells in either of the two battery compartments. Each of the two pairs of starboard and two pairs of port tanks is equipped with capacity gages accessible from the battery spaces.

D. GALLEY EQUIPMENT

9D1. Galley and scullery sinks. The officers' pantry, the galley, and the crew's mess room are provided with sinks. Each is supplied with hot and cold water. The sink drains are connected to the sanitary drainage system.

9D2. Coffee urn. The crew's mess room is provided with a 5-gallon electrically heated coffee urn with a tap for drawing coffee in the mess room. A cold water line supplies the urn with water.

9D3. Scuttlebutt. The main drinking water dispensing equipment aboard the submarine is the scuttlebutt, or the drinking fountain. One scuttlebutt is located in the officers' pantry, and one in the crew's mess room. Each scuttlebutt is provided with a cold water supply line and drain to the sanitary tank drainage system. Before the water enters the scuttlebutt in the mess room, it is passed through a cooling coil located in the cool room. The water for the scuttlebutt in the officers' pantry is cooled by the small refrigerator in the pantry.

9D4. Lavatories and showers. Each lavatory is provided with one cold water line, a hot water line, and a drain to the sanitary tank. There is one lavatory in the forward torpedo room, two in the officers' quarters, one in the commanding officer's stateroom, one in the chief petty officers' quarters, two in the crew's quarters, and one in the after torpedo room.

The showers are provided with hot and cold water. The deck drain for each shower is connected with the sanitary tank drainage system. There is one shower in the starboard forward corner of the officers' quarters and two showers in the after end of the crew's quarters.

E. PLUMBING

9E1. Sanitary drainage. All lavatories, sinks, showers, scuttlebutts, and heads torpedo room lavatory, the refrigerator, and the pantry sink.

drain into No. 1 and No. 2 sanitary tanks through the sanitary drainage system consisting of the sanitary drainage piping and valves. The No. 1 sanitary tank, located inside the MBT No. 1 has two sanitary drainage mains connecting to it, one on the starboard and one on the port side. The starboard main to No. 1 sanitary tank receives the drainage from the commanding officer's lavatory, wardroom, stateroom No. 1 lavatories, and the officers' shower. The port drain receives discharge from the chief petty officers' lavatory, wardroom, stateroom No. 2 lavatory, the forward

The No. 2 sanitary tank, located in the after starboard end in the after battery compartment, receives the drainage from the sanitary drain which collects the discharge from the following: the galley sink, the scullery sink, the scuttlebutt, the crew's lavatories, the shower, and the washroom decks. The officers' head in the forward torpedo room empties directly into the No. 1 sanitary tank; the after head in the crew's quarters empties into the No. 2 sanitary tank. The forward head in the crew's quarters and the head in the maneuvering room discharge directly to the sea.

F. HEADS

9F1. Expulsion type head. There are two air expulsion type water closets (heads), each

Figure 9-3. Expulsion type head.

fitted with an auxiliary hand pump, one in the crew's quarters, and one in the after end of the maneuvering room. (See Figure 9-3.)

The water closet installation consists of a toilet bowl over an expulsion chamber with a lever and pedal controlled flapper valve between, which is weighted to hold water in the toilet bowl and seats with pressure of the expulsion chamber.

Each installation operates as a separate unit with its own flood, blow, and discharge lines. The toilet bowl is provided with a sea flood with stop and sea valves. The expulsion chamber has a discharge line with swing check, gate, and plug cock valves. The blow line to the expulsion chamber receives air through a special rocker valve which, when rocked in one direction, admits air from the low-pressure air service line into a small volume tank until a pressure of approximately 10 pounds above sea pressure is reached. When rocked in the opposite direction, the rocker valve directs the volume of air into the expulsion chamber. A sea pressure gage, a volume tank pressure gage, and an instruction plate are conveniently located.

Before using a water closet, first inspect the installation. All valves should have been left shut. Operate the bowl flapper valve to ascertain that the expulsion chamber is empty.

Shut the bowl flapper valve, flood the bowl with sea water through the sea and stop valves, and then shut both valves. After using the toilet, operate the flapper valve to empty the contents of the bowl into the expulsion chamber, then shut the flapper valve. Charge the volume tank until the pressure is 10 pounds higher than the sea pressure. Open the gate and plug valves on the discharge line and operate the rocker valve to discharge the contents of the expulsion chamber overboard.

Figure 9-4. Gravity flush type head.

Shut the discharge line valves and leave the bowl flapper valve seated. For pump expulsion, proceed as previously stated except that the contents of the waste receiver are to be pumped out after the gate and plug valves on the discharge line have been opened.

If, upon first inspection, the expulsion chamber is found flooded, discharge the contents overboard before using the toilet. Improper operation of toilet valves should be corrected and leaky valves overhauled at the first opportunity.

9F2. Gravity flush type head. There are two gravity flush type water closets (heads), one in the forward torpedo room for the officers, and one in the after end of the crew's quarters. (See Figure 9-4.)

The water closet installation consists of a toilet bowl over a waste receiver with a lever and pedal-controlled flapper valve between, which is weighted to hold water in the toilet bowl and seats with the pressure in the tank.

Each toilet bowl is provided with a flood line with stop and sea valves. The water closets are located over the sanitary tanks and discharge directly into them.

Before using a water closet, first inspect the installation. All valves should have been left shut. Operate the bowl flapper valve to ascertain that the waste receiver is empty. Shut the bowl flapper valve, flood the bowl with sea water through the sea and stop valves, and then shut both valves. After using the toilet, operate the flapper valve to empty the contents of the bowl into the waste receiver and sanitary tank.

G. DISTILLATION

9G1. Submarine distilling equipment. The distillers in use on modern submarines are either the Kleinschmidt Model S, or the Badger Model X-1. Two stills are installed on all later class submarines. The Kleinschmidt model is discussed and illustrated in this text. (See Figure 9-5.)

9G2. Consumption of fresh water. A modern submarine during a war patrol will consume on the average approximately 500 gallons of fresh water per day for cooking, drinking, washing, and engine make-up water. In addition to this consumption, the main storage batteries require about 500 gallons

of battery water per week; giving a total requirement of at least 4000 gallons per week. This minimum requirement will allow each man in the crew to have a bath at least twice a week.

9G3. Fresh water stowage capacity. The normal fresh water stowage capacity is about 5400 gallons; 1200 gallons of this is battery water and is stored in the battery water tanks. This water will last only about 10 days and it is good practice not to allow the fresh water on hand to drop below one-half the normal capacity. The area of operations is usually the determining factor as to when the distillers can be used.

9G4. Principles of distilling action. The knowledge of distilling liquids

The distiller can be supplied with feed water from the main engine salt water circulating pump sea suction, from the main motor circulating water system, and from the ship's fresh water system. The latter feed is used when redistilling ship's fresh water for battery use.

Part of the sea water flows out of the still as distilled water, and collects in the distillate tank. It can be transferred by blowing to the fresh water system or to the battery water system for stowage.

comes from ancient days. Distillation is simply the boiling of a liquid and the condensing of its vapor to the liquid state again. In the boiling, much or all of any impurities or undesired contents are left behind, so that the condensed liquid is free of them. If a teaspoon is held in a cloud of steam arising from a teakettle, the vapor will condense on the spoon and the resulting liquid is distilled water.

9G5. The purifying action in distilling sea water. In sea water, salt and other substances are dissolved or held in solution. Sea water does not boil at the same temperature, 212 degrees F, as does fresh water, but at a temperature a few degrees higher. When the sea water boils, it is only the water (H2O) that is vaporizing at this temperature, and if that pure vapor is led to another clean container where it may condense, the result: is pure distilled water. The salt (sodium chloride) and other solid ingredients of the sea do not vaporize and hence do not come over into the distilled water.

9G6. Brief explanation. A brief explanation is given of the actions that take place in the Kleinschmidt still, without mentioning mechanical details, in order that these actions may be easily understood.

The distilling process in the Kleinschmidt still is continuous, with sea water being supplied at the rate of about a gallon per minute.

Figure 9.5. Kleinschmidt distillers.

The desuperheating tank, the purpose of which is to supply the cooling water to the still and to lubricate the lobes of the compressor at the top of the still, can be replenished from the distillate tank. The remaining sea water is concentrated brine and flows out separately.

Inside a cone-like casing, a long length of tubing is coiled. This casing is set with

the small end down. Actually there are ten such cones, nested together. Cold sea water enters at the bottom between

temperature lower than its new condensation point, and so is able to condense. This type of apparatus is

the cones; that is, it flows around the outside of the tubing. Here, on its way up, it is heated, so that it is boiling when it emerges from between the cones at the upper end. The vapor is led through a vapor separator into a compressor, where it is compressed, and is then discharged down into the *inside* of the tubing. On the way down through the tubing, this vapor is gradually cooled by contact with the colder tubing walls, finally condensing therein and flowing out as pure distilled water to a storage tank. The nested cones of tubing, therefore, act as a heat exchanger. The distilled water is technically known as *distillate* or *condensate*.

9G7. Necessity of compressing the vapor.
A question may arise as to why the vapor is compressed in the still. The explanation involves several considerations:

The conical crest of tubes serves three purposes: 1) vaporization of the feed water, 2) condensation of the vapor, and 3) cooling of the hot condensed liquid to a lower temperature. In the lower part of the nest, the feed water is at the temperature of sea water; the temperature increases during the upward flow, so that the feed water leaves the nest boiling. The feed water in the upper part of the nest is therefore very hot. On the downward flow, the vapor is condensed in the upper part of the nest, and in the lower part of the nest, the hot condensed liquid is cooled.

Sea water does not boil at the same temperature, for a given pressure, as does fresh water, but at several degrees higher. The feed water in the upper part of the nest is, therefore, actually above

accordingly called a *vapor compression still.*

9G8. Heat input of still.
Compression of the vapor serves another purpose also. On starting operation of the still, the feed water is brought up to boiling temperature by the electric heaters. After the still is in normal operation, there will be a steady heat loss of definite amount through the insulation and in the outgoing condensate and brine overflow. This heat loss is balanced by an input of energy from the electric motor, which is transformed to heat by the compression of the vapor. Theoretically, this input of heat by the compressor maintains the heat balance at a constant level, and it is possible to operate the still with all electric heaters turned off. In actual practice, however, most of the heaters are usually left on after the still is in normal operation.

9G9. Vent to atmosphere.
Since the boiling of the sea water takes place inside the shell of the still, it is necessary to prevent any increase of pressure on the boiling water, for increased pressure here would raise the boiling point and put the whole system out of balance and probably out of operation. The situation is different in the compressor; when the vapor goes into the compressor, it is sealed off from the boiling liquid and may then be compressed without affecting the boiling point. Therefore, in order that the boiling may always take place at atmospheric pressure as found within the submarine, a pipe called the *vent* leads from the *vapor separator* and out through the bottom of the still. This vent, being open to the atmosphere at its end, insures that the pressure in the vapor separator is always at the pressure of the surrounding atmosphere. A distant

212 degrees F. But the vapor from the boiling water is no longer sea water; it is fresh water vapor, and fresh water vapor at atmospheric pressure can condense only at 212 degrees F. When a vapor is compressed, its boiling point, or its condensation point, rises. By compressing the vapors in the still, its condensation point is raised above the temperature of the hot feed water in the upper part of the nest. Therefore, when the compressed vapor enters the nest, it finds a

reading dial thermometer is connected by a flexible tubing to the vent to give the temperature in the vent pipe.

Although this open vent pipe leads downward out of the still, the steam when in normal amount inside will not flow out, because of the pressure of the outer atmosphere

through the vent. Actually, the interior and exterior pressures are so maintained that there is a very small excess of pressure inside the still. This causes a slight feather of steam to appear at the vent, which is an indication that the still is operating satisfactorily. Any excess steam that the compressor cannot handle, however, will be able to pass out through the vent, which thus acts as a safety device.

The vent pipe also serves to permit drainage to the bilge of any slight amount of liquid carried into the vapor by the violent boiling action, and prevents it from gathering on the floor of the vapor separator. A small drain tube leading down from the compressor to just above the vent pipe in the vapor separator also permits water and oil to drain from the compressor seal out through the vent.

9G10. Portion of sea water not distilled. All of the incoming sea water cannot be distilled, for some must remain undistilled to carry away the concentrated salt content left from the

height as to insure that the interior overflow heat exchanger is always full of liquid, and therefore always exerts its full heating effect on the sea water inside.

9G12. Time required to start sea water boiling in still. When starting the still, from 60 to 90 minutes are required to bring the temperature up to the boiling point.

9G13. Heat balance in the still. It may be interesting to indicate the heat flow through

distilled portion. This undistilled portion, which is concentrated brine, is maintained at a level just above the top coil of the heat exchanger by overflow pipes. It flows down through these overflow pipes into a separate conical passage, called the *overflow heat exchanger*, located around the nest of tubes, where it gives up some of its heat by conduction through the metal walls, thus helping to heat the incoming feed water.

9G11. Overflow weir cup. The overflow pipe, after leading out of the overflow heat exchanger at the bottom of the still casing, rises again a short distance. At the top of the upright overflow pipe, the brine flows out through an opening called the *weir*, which meters or measures the quantity of brine overflow in gallons per hour. The overflow brine passing out of the weir falls into an open cup and then drains down to a storage tank called the *brine receiver*, from which it is discharged to the sea.

Since water in any U-shaped container must always be at the same level in both arms of the U, the open weir is located at such a

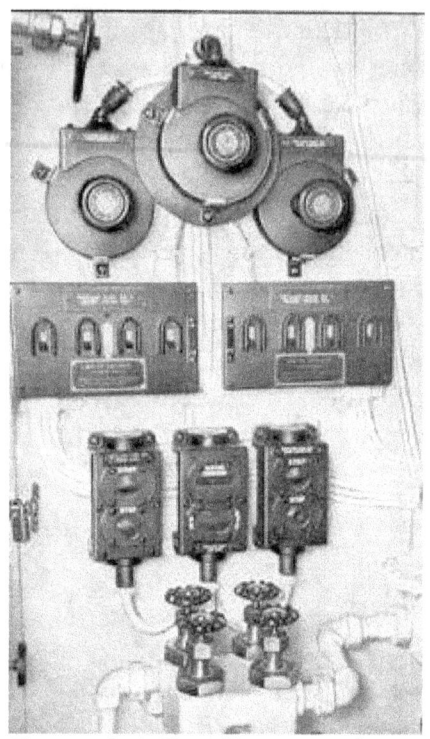

Figure 9-6. Distiller controls.

the various parts of the still in actual quantities. The following is an example:

Heat input: 10,185 Btu per hour come in through the compressor; 5,940 Btu per hour come in through the electric heaters. This is after the still is fully operating. The total input of heat is 16,125 Btu per hour.

Heat loss: This total quantity of heat flows out through four separate paths, as follows:

1,825 Btu per hour in condensate, 11,600 Btu per hour in overflow, 400 Btu per hour from vent, 2,300 Btu per hour by radiation from hot metal parts, or a total heat loss of 16,125 Btu per hour. The heat balance is not always at this exact number of Btu per hour, for

separate stills (Figure 9-6) each with its necessary control devices.

Two stills are necessary, not only as a safety factor, but also to provide sufficient distilled water. These stills may normally be run 300 to 350 operating hours without cleaning. Each gives 40 gallons per hour. This means a total of 24,000 to 28,000 gallons of distilled water. The consumption of distilled water

various momentary changes or rate of feed and temperature of sea water, voltage fluctuations in motor, or other operating conditions, will naturally cause it to vary around any average number. This heat balance of the still is very sensitive and all changes that may be necessary in the operation conditions should be made slowly.

9G14. Purity of the distilled water. If no leaks are present, the distilled water will contain only about one part of salt to a million parts of water. The distiller cannot, of course, remove any volatile liquids; that is, liquids that boil at or below the boiling temperature of water. For example, in badly polluted harbors or streams, a trace of ammonia may be present in the distilled water; and in improperly chlorinated waters, a trace of chlorine may likewise come over in the distilled water.

9G15. Two-still system. In the complete submarine distilling system, there are two

is about 600 gallons per day for all purposes. On a war patrol lasting 60 days, the total consumption will be about 36,000 gallons, and may run higher in the tropics.

9G16. Water for the storage batteries. Distilled sea water is fit for human consumption and for the storage batteries. It may happen that fresh water is taken on board from some shore source. Such fresh water is not suitable for storage battery use until it has been distilled. Fresh water taken aboard in any foreign port should always be boiled or distilled before use. Only fresh water definitely known to be pure may be used without distilling or boiling for drinking, cooking, or personal use. In distilling fresh water that is taken aboard, the operation of the still is practically the same as when distilling sea water, the difference being that the overflow is returned to the ship's fresh water tanks from the brine tank instead of being discharged overboard.

107

10
TRIM AND DRAIN SYSTEMS

A. TRIM SYSTEM

10A1. General description. As explained in Chapter 5, the balance and stability of the submarine can be upset by an unequal distribution of weights in the ship. The trim system is employed chiefly to correct this condition by regulating the quantity of water in the variable tanks.

Figure A-12 illustrates the general

lines; all of these tanks make up the variable ballast tanks group. The remaining flood and suction lines are connected to the negative and the safety tanks, called the special ballast tanks.

Cross connection of the trim pump and the drain pump is made by two flanged connections on the after end of the longitudinal axis of the manifold. One

arrangement of the trim system in the submarine. It shows the trim pump manifold, the main flood and suction lines, the valves, and the connections to the various trim system tanks.

The trim manifold, located on the port side aft in the control room, is the center of control for the entire system in that it directs the flow of water to the various tanks. It is a casting divided into two longitudinal compartments known as the suction and discharge sides. The discharge side of the manifold contains eight discharge control valves. One of these valves is the trim pump discharge valve and connects the discharge side of the manifold with the discharge side of the trim pump. The suction side of the manifold contains eight suction control valves, and is connected to the suction side of the pump through the trim pump suction valve.

The remaining seven discharge and seven suction valves control the flood and suction from the following lines:

1. Trim pump suction and overboard discharge line.
2. Trim line forward flood and suction.
3. Trim line aft flood and suction.
4. Auxiliary ballast tank No. 1 flood and suction.
5. Auxiliary ballast tank No. 2 flood and suction.
6. Negative tank flood and suction.
7. Safety tank flood and suction.

The trim lines forward and aft serve the two trim tanks and the two WRT tanks, while auxiliary ballast tanks No. 1 and No. 2 are served by their own flood and

connection is on the discharge side, the other on the suction side.

The trim pump, located in the after end of the pump room, provides pumping power for the system. It draws water into its suction side, through the suction side of the manifold, from the tank being pumped, and discharges it through its discharge side into the discharge side of the manifold, which directs the water to the tank being filled. When it is desired to pump to one of these tanks by means of the trim pump, the discharge valve on the trim pump manifold controlling this particular tank is opened. When water is to be removed from a tank by means of the trim pump, its valve on the suction side of the manifold is opened. Thus, the trim manifold control valves serve to put any part of the trim system on suction or discharge. For example, in pumping from forward trim tank to after trim tank, the water is drawn through lines from the forward trim tank through the suction side of the manifold and into the suction side of the trim pump, and forced by pump action through the discharge side of the trim pump, through the discharge side of the trim manifold, and then through lines into the after trim tank. The trim line forward is a 3-inch line extending from the trim manifold in the forward torpedo room. The forward trim manifold controls the flooding and pumping of the forward trim tank and the forward WRT tank.

The trim line aft is also a 3-inch line, terminating in the after torpedo room at the after trim manifold, which controls the flooding

suction

and pumping of the after trim tank and the after WRT tank.

The Nos. 1 and 2 auxiliary ballast tanks are piped directly to their suction and discharge valves on the trim pump manifold. The flooding or pumping of these tanks can be accomplished only through the trim manifold. On the other hand, the flooding and draining of the safety and the negative tanks can be accomplished in two ways: either by the use of their suction and discharge valves on the trim manifold, or directly from sea by use of their flood valves. In the latter case, the blowing is accomplished by opening the flood valves and admitting compressed air into the tanks, thus forcing the water out; while the tanks are flooded by opening both the flood and the vent valves, allowing the sea to enter directly into the tanks.

The trim pump suction and overboard discharge line, connecting the trim manifold with the sea, provides the trim system with an overboard discharge to, or direct flooding from, the sea. In addition to the suction and discharge valves on the trim manifold, this line has also a sea stop valve and a magazine flood valve. The sea stop valve is used to shut off the sea from the trim system and the magazine flood valve. The magazine flood valve guarantees, when the sea stop valve is open, an immediate source of sea water to the ammunition stowage and the pyrotechnic locker.

As stated before, the main function of

water ballast from one variable tank to another, adding water to the variable tanks or discharging excess water from the tanks overboard. Therefore, the water handled by the trim system is measured in pounds; and a gage, graduated in pounds to show the amount of water transferred by the trim pump, is located above the trim manifold where the operator can observe its reading.

Because the trim pump used on the latest fleet type submarine is of the centrifugal type, it must be primed before beginning the operation. A priming pump is used for this purpose. It primes the trim pump by removing all air from the trim pump casing, the trim manifold, and the lines leading to it, thus allowing water to replace the air in this equipment and fill it completely. It should be noted that some submarines are equipped with a trim pump of the reciprocating or plunger type, similar to the drain pump shown in Figure 10-5.

The trim system can also be used to supply or drain water from the torpedo tubes. Water for flooding torpedo tubes is normally taken from the WRT tanks through tube flood and drain lines. These lines are controlled by the torpedo tube flood and drain valves.

The trim line forward and the trim line aft are provided with hose connections, one in each compartment of the submarine. These connections can be used for fire fighting, or for bilge suctions in those compartments that do not have bilge suction facilities. Of course, if the connections are used for

the trim system is to shift and adjust the distribution of weight throughout the submarine. This is done by means of transferring

bilge suction the trim line must be on suction, and if for fire fighting, the line must be on discharge.

B. TRIM PUMP

10B1. Source of power. The trim pump, Figure 10-1, located on the port side of the pump room just forward of the after bulkhead, is driven by a 10/25 horsepower motor directly connected by means of a flexible coupling to the drive shaft of the trim pump.

The controller relay panel for the motor is mounted on the after bulkhead of the pump

room. However, the motor is started or stopped by push button controls in the control room. Once started by these controls, the speed of the pump and, thereby, the rate at which water is moved in the system, is regulated by a rheostat control, also located in the control room just below the push button switch (Figure 10-2).

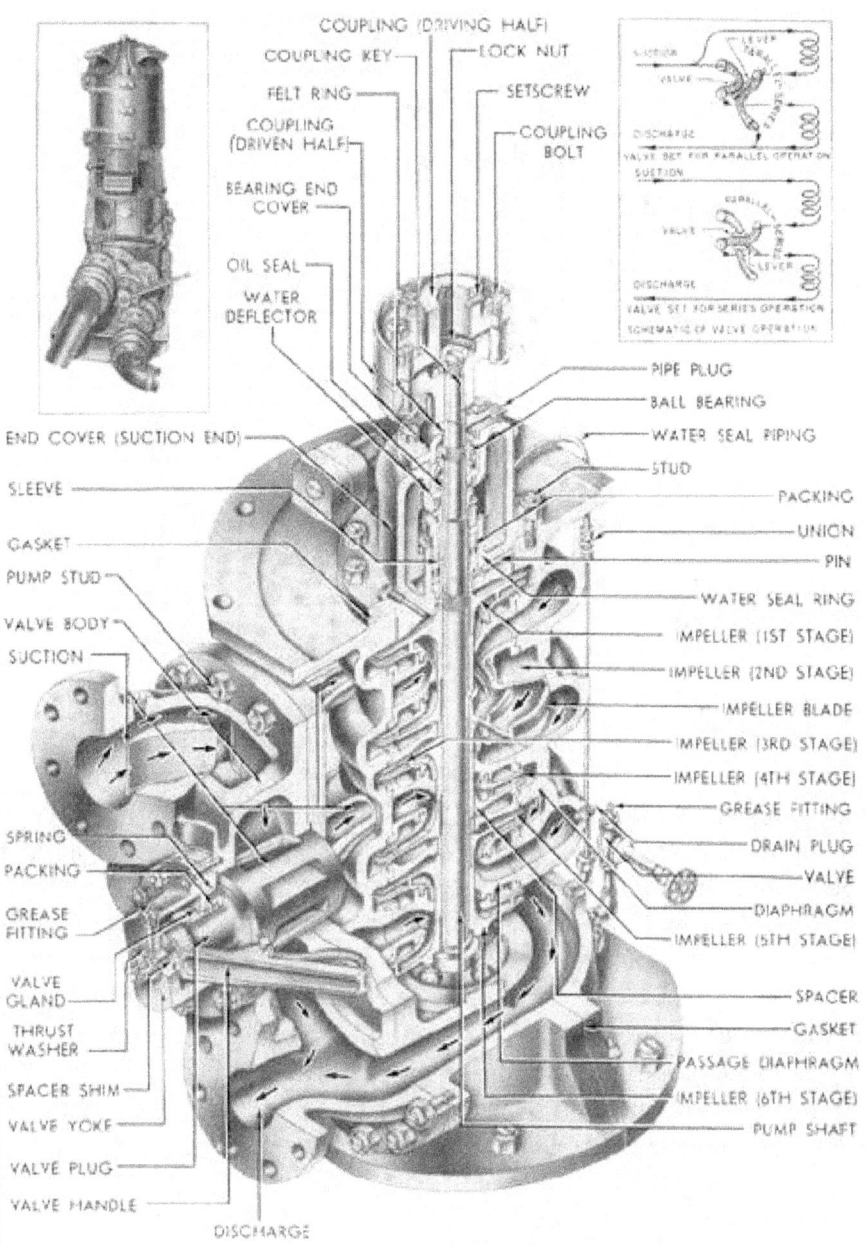

COUPLING (DRIVING HALF)
COUPLING KEY
LOCK NUT
FELT RING
SETSCREW
COUPLING (DRIVEN HALF)
COUPLING BOLT
BEARING END COVER
OIL SEAL
WATER DEFLECTOR

PIPE PLUG
BALL BEARING
WATER SEAL PIPING
STUD
END COVER (SUCTION END)
SLEEVE
PACKING
GASKET
UNION
PUMP STUD
PIN
VALVE BODY
WATER SEAL RING
SUCTION
IMPELLER (1ST STAGE)
IMPELLER (2ND STAGE)
IMPELLER BLADE
IMPELLER (3RD STAGE)
IMPELLER (4TH STAGE)
GREASE FITTING
SPRING
DRAIN PLUG
PACKING
VALVE
GREASE FITTING
DIAPHRAGM
IMPELLER (5TH STAGE)
VALVE GLAND
SPACER
THRUST WASHER
GASKET
PASSAGE DIAPHRAGM
SPACER SHIM
IMPELLER (6TH STAGE)
VALVE YOKE
PUMP SHAFT
VALVE PLUG
VALVE HANDLE
DISCHARGE

Figure 10-1. Trim pump.

110

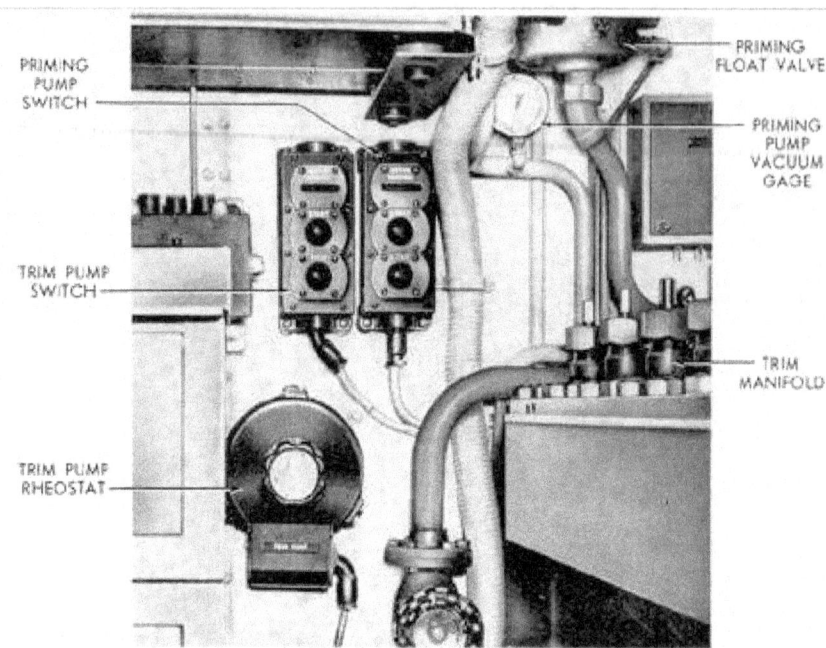

Figure 10-2. Trim pump controls.

While the trim pump is driven by an electrical motor, the starting of the motor does not guarantee that the trim pump will pump water, for the trim pump, being of the centrifugal type, cannot pump air. Therefore, it cannot be operated until the system is free of air.

10B2. Priming pump. Freeing the system of air is the purpose of the priming pump, located outboard of the trim pump. Since any appreciable amount of air entering the inlet side of the trim pump will cause it to lose suction and thereafter run without pumping, it is necessary to know when this critical condition has been reached. To indicate the amount of air in the trim system a vacuum gage is provided which is mounted in the control room. When the vacuum gage indicates less than 21 inches of vacuum, the priming pump must be operated before the trim pump is started. The priming pump, like the trim pump, is started by push button controls located in the control room. The

priming pump is a vacuum pump with a float valve in the line, running from the priming pump to the trim manifold and the trim pump casing. The valve consists of a float with a ball-ended stem. The purpose of the float is to permit the passage of air and to prevent the passage of water into the priming pump. As the water rises in the float valve, the upper part of the ball-ended stem is automatically forced against the valve seat, thus preventing the sea water from entering the priming pump.

The priming pump is a self-priming centrifugal displacement pump consisting of three major parts: rotor, lobe, and port plate. The rotor is made up of a series of curved plates projecting radially from the hub. The lobe is elliptical in shape and forms the outer casing for the rotor. The port plate consists of two inlet and two outlet ports corresponding to the inlet and outlet ports on the rotor. The pump is end-mounted on the direct driving electric motor as shown in Figure 10-3.

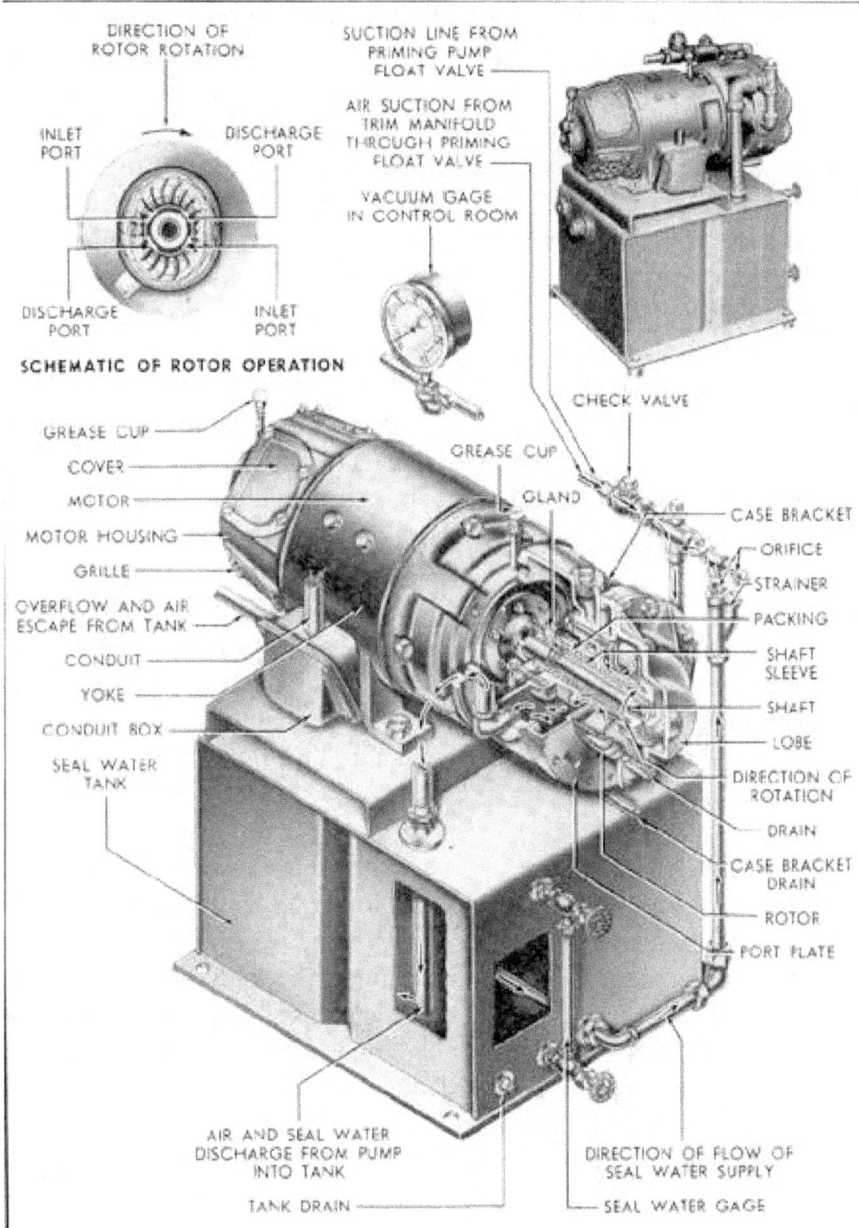

Figure 10-3. Priming pump.

Before starting the priming pump, it is necessary first to provide seal water to it. This water is needed to fill the lobe partly and provide a water seal. Water obtained from the seal water tank should be added until the seal water

The periphery of the impeller is open, as shown in Figure 10-1.

In operation, water enters the eye of the impeller, is picked up by the vanes and accelerated to a high velocity by the

gage shows two-thirds full. (See Figure 10-3.) Serious damage may result if the pump is allowed to run in a dry condition. The motor is then started by the push button control in the control room.

In operation, the rotor revolves in the lobe, which has been partially filled with water, at a speed high enough to throw the water out from the hub by centrifugal force. This results in a solid elliptical-shaped ring of water revolving at the same speed as the rotor. In Figure 10-3, showing rotor operation, it may be seen that a ring of water for a given rotor section, guided by the lobe, will move in and out from the hub, forming a liquid piston. As the rotor passes the inlet port, the water ring is farthest from the hub, and air is permitted to enter. As the rotor advances to discharge port, the air space becomes less and air is forced out the discharge port. This cycle is repeated twice for each revolution of the rotor. When the vacuum gage registers 21 inches of vacuum in the trim system, the priming pump can be stopped and the trim pump started.

10B3. Operation of the trim pump. A brief review of the general principles of centrifugal pumps will be helpful in understanding the operation of the trim pump. As the name implies, this type of pump employs centrifugal force to move a liquid from a lower to a higher level. In its simplest form, the centrifugal pump consists of an impeller rotating in a watertight casing provided with inlet and outlet ports.

The impeller consists of two parallel disks with curved vanes or bulkheads radiating from the hubs and between

rotation of the impeller, and discharged by centrifugal force into the casing and out of the discharge port. When water is forced away from the eye of the impeller, a suction is created and more water flows in. Consequently there is a constant flow of water through the pump. An air bubble in the inlet port of the pump will interrupt the action of the pump since it will, upon entering the impeller, break the suction at the eye which is dependent on the presence of water. For this reason, the pump casing and the system served by the pump must be solidly filled with water before pumping is commenced. This is the function of the priming pump. The centrifugal pump described above has only one impeller and is known as a single-stage pump; a pump with four impellers would be known as a four-stage pump; with six impellers, a six-stage, and so on. In practice however, any pump with more than one stage is referred to as a multi-stage pump.

The mechanical details of the trim pump are shown in Figure 10-1. It will be seen that it is a six-stage centrifugal pump. The valve shown on the forward end permits either parallel or series operation and is manually operated. The schematic diagram in the upper right corner of the illustration shows the flow of the water being pumped for both series and parallel operation. With the manually operated series-parallel valve in the SERIES position, the incoming water enters the first stage, proceeds through the second and third stages, and then back through the series-parallel valve to the fourth, fifth, and sixth stages. With the series-parallel valve in the parallel position, half of the inlet water proceeds through the first, second, and third stages, and is then discharged through the series-parallel valve. Simultaneously, the other half of

the disks. One of these two disks (upper or lower, depending upon where the water is brought in) has an inlet port or circular opening called the eye, concentric with the hub of the impeller. Actually then, one disk holds the impeller to the shaft while the other admits the water.

and sixth stages and is then discharged directly. Series operation of the pump produces twice the discharge pressure, but only one half the volume produced by parallel operation. The pump is operated in series only when the submarine is at depths of approximately 250 feet or more and discharging to the sea, the higher pressure being necessary to overcome the greater sea pressure encountered at that depth.

the inlet water is directed by the series-parallel valve to the fourth, fifth,

not be started if this gage registers less than 21 inches of vacuum. If the gage shows less than 21 inches, the system must be restored to the proper condition by using the priming pump. When the gage registers the required 21 inches, the priming pump is stopped and the trim pump started.

The trim pump should not be operated at speeds greater than are necessary to produce a rate of flow specified for a given depth. The accompanying table lists the valve

DEPTH	PUMP OUTPUT	VALVE POSITION
On surface	1,500-2,500 pounds per minute	Parallel
0-200 ft.	1,500 pounds per minute	Parallel
200-250 ft.	1,250 pounds per minute	Parallel
400 ft. or deeper	1,000 pounds per minute	Series
250-400 ft.	1,000 pounds per minute	Series

To summarize, it must be remembered that before starting the trim pump, it is necessary to make certain that the trim system lines and the pump casing are free of air, as explained earlier in this section. A vacuum gage is provided to indicate the condition existing in the system. The trim pump must

position and pump output in pounds of water per minute, recommended at different depth levels.

The pump should not be operated at a motor speed greater than 2,400 rpm. Excess speeds place an overload on the bearing and mechanical parts of the pump and may cause a breakdown.

C. MANIFOLDS

10C1. Trim manifold. In section A of   board mounted directly above it.

this chapter, the trim manifold is referred to as the center of distribution for the trim system. It acts as a switchboard between the trim pump and the lines of the system, providing a centralized station to direct the flow of water to and from the variable tanks. Used in connection with the trim manifold, but connected to each variable tank, is a measuring gage, or liquidometer. These gages record the amount of water in each tank, and provide the diving officer with an indication of the amount of water ballast being redistributed by the trim manifold through the trim system. The trim manifold is mounted hip-high of the port side of the control room just forward of the after bulkhead, with the gage

Figure 10-4 shows the mechanical construction of the trim manifold, and also the proper nomenclature of its details.

It will be seen that the manifold is a box-like, two-piece casting divided internally into two longitudinal compartments, known respectively as the suction and discharge sides. The suction side contains eight suction control valves, while the discharge side has eight discharge (or flood) control valves. Each of these 16 valves is of the dish and seat type, with rising stems and individual bolted-on bonnets. Name plates attached to each bonnet indicate the function of that particular valve.

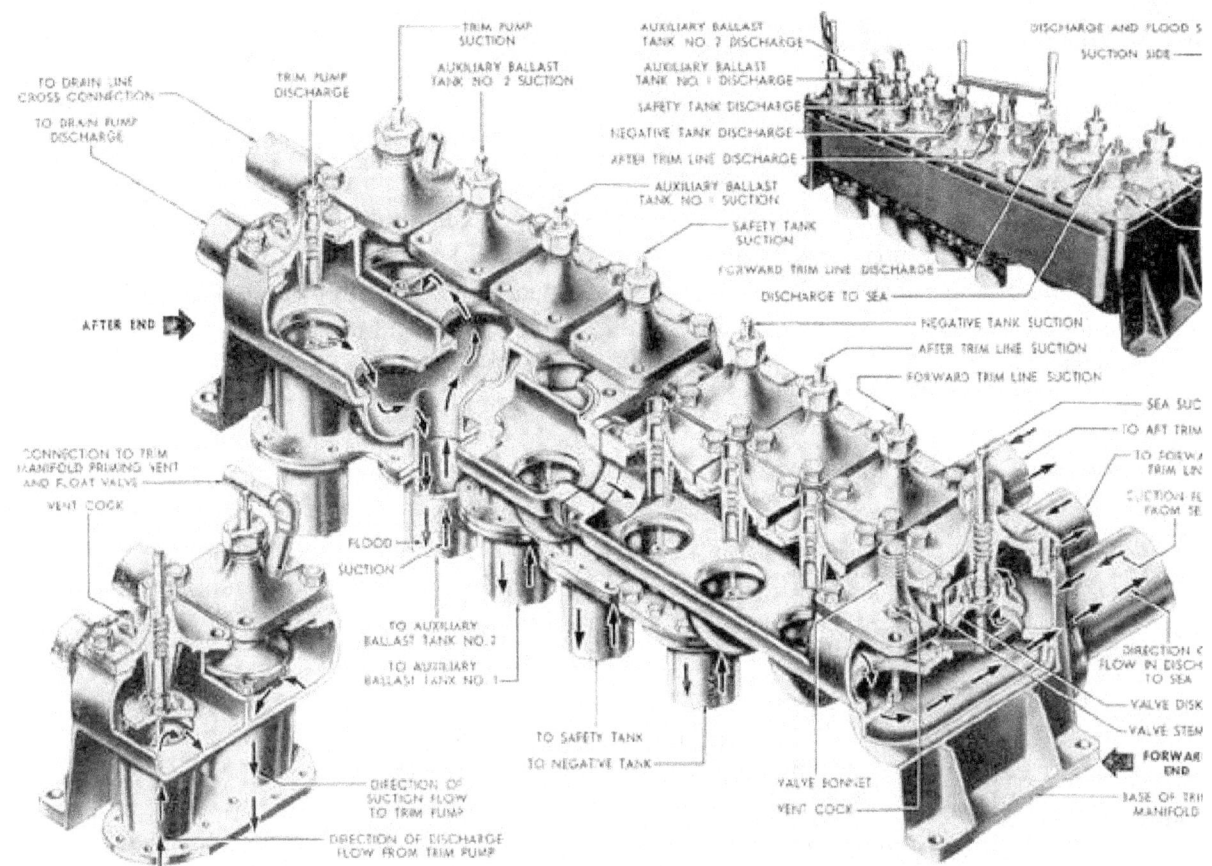

Figure 10-4. Trim manifold.

Starting from the after end outboard of the trim manifold, the valves in the suction and discharge sides control the following components:

OUTBOARD (Suction)	INBOARD (Discharge)
1. Trim pump suction	9. Trim pump discharge
2. Auxiliary ballast tank No. 2 suction	10. Auxiliary ballast tank No. 2 discharge
3. Auxiliary ballast tank No. 1 suction	11. Auxiliary ballast tank No. 1 discharge
4. Safety tank suction	12. Safety tank discharge
5. Negative tank suction	13. Negative tank discharge
6. After trim line suction	14. After trim line discharge
7. Forward trim line suction	15. Forward trim line discharge
8. Sea suction	16. Discharge to sea

The discharge valves are all on the starboard side of the manifold with the corresponding suction valves opposite them on the port side. A special forward of the torpedo tubes. (See Figure A-12.)

The forward and the after WRT and trim

wrench is provided for operating the valves.

Flanged outlets are cast integral with the manifold to connect with the lines of the system. Two outlets on the after end lead to the drain line cross connection and to the drain pump discharge to permit emergency use of the drain pump to actuate the trim system.

In all pumping operations, the trim pump suction and the trim pump discharge valves on the manifold must be opened to permit flow within the system. To flood a tank, the discharge valve for that tank must be opened at the trim manifold; to pump a tank, its suction valve must be opened. This should be done before the trim pump is started. All valves on the manifold should be shut immediately after the pumping operation is complete. Figure 10-4 shows the direction of flow when flooding or pumping auxiliary ballast tank No. 2.

10C2. Forward and after WRT and trim tank manifold. The WRT and trim tank manifolds are used in conjunction with the trim manifold to control the flooding and pumping of thre WRT tanks and the trim tanks, both fore and aft.

The forward trim manifold is located in the forward torpedo room, portside, aft, of the torpedo tubes. The after trim manifold is located in the after torpedo room, portside,

tank manifolds are identical in operation and construction, differing only in the fact that they serve different tanks.

The body of each trim manifold is a two chambered casting containing two valves which control flood and suction of the WRT tank and the trim tank respectively. The after valve in the after torpedo room and the forward valve in the forward torpedo room control the trim tanks. The valves are of the disk and seat type with bolted bonnets. The connecting passage between chambers of the integrally cast valve casting allows either valve to be operated independently. The handwheels carry name plates designating the uses of the individual valves.

When open, the manifold valve marked trim tank flood and suction permits the flooding or pumping of the trim tank from or into the trim system when the trim line is on service.

The other valve, marked WRT tank flood and suction, permits the flooding or pumping of the WRT tank from or into the trim system when the torpedo tube drain stop valve to the WRT tank is open.

10C3. Torpedo tube drain manifold. In Section A of this chapter, the flooding and draining of the torpedo tubes were mentioned as functions of the trim system. These functions are controlled by the torpedo tube drain manifolds. Two of these manifolds are located in

the forward torpedo room, each servicing three torpedo tubes; and two

attached to the back of the casting. Separate control levers and connections in the after torpedo room, each servicing two torpedo tubes.

The body of the torpedo tube drain manifold is a three-chambered casting, housing three cam-actuated plunger type valves, and provided with flanged outlets for connection to the trim system and to the torpedo tube drains. The cam mechanisms are

are provided for each of the valves.

Each, hand lever operates one cam through the action of its connecting rod and cam lever. In draining or flooding the tubes, the manifold valves are used in conjunction with the torpedo tube drain stop valve to the WRT tank which must be open when draining from the tubes to the WRT tank.

D. VALVES

10D1. Trim pump sea stop valve. When it is desired to discharge water ballast from any part of the trim system to sea, the trim pump sea stop valve must be opened, thus providing a passage from the trim manifold through the pressure and outer hulls to the sea. The same line is used to permit water to enter the system from the sea when additional water ballast is to be added. This valve is located on the port side of the control room, directly below the trim manifold. (See FigureA-12.)

10D2. Torpedo tube drain stop valve to the WRT tank. The torpedo tube drain stop valve to the WRT tank serves as a stop valve between the WRT tank and the individual torpedo tube drain valves.

There is a torpedo tube drain stop valve

to the WRT tank in both the forward and the after torpedo rooms. Both of these valves are identical in function and construction.

10D3. Magazine flood valve and testing casting. The magazine flood valve and testing casting provide an emergency method of flooding the magazine compartment.

The magazine flood valve is used to control this emergency flooding system. The testing casting is used to check the magazine flood valve to make certain that it is ready for immediate use. Both the magazine flood valve and the testing casting are located in the control room on the magazine flood line of the trim system. The accessory box, containing the operating plug and wrench, is mounted directly above the testing casting.

E. DRAIN SYSTEM

10E1. Functions. In submarines, as in all ships, a certain amount of water accumulates inside the hull from various sources. The most important of

with the operation of the submarine. This water is pumped out by the drain system which consists essentially of the drain pump and the piping connecting

these sources are:

a. Leakage at glands around propeller shafts, Pitometer log, sound gear, periscopes, and similar equipment.

b. Draining of air flasks, manifold drain pans, conning tower deck gun access trunk, and escape trunk.

c. Condensation from air-conditioning cooling coils.

This water drains off into the bilges and wells where a number of bilge sumps with strainers are provided, from which the bilge water can be pumped.

The bilge sumps and wells are pumped periodically to prevent the excess free water from overflowing the bilges and interfering

the pump with the sumps and other drainage points in the submarine. The general arrangement shown in FigureA-12 is used in the following functional description:

The drain pump located in the pump room provides the suction for the drain system. The pump is started and stopped by means of an electric push button switch located nearby in the pump room. The drain pump has a suction and a discharge connection. A suction line equipped with a strainer and a sight glass connects the suction side of the pump with the main forward and after drain lines, called the *drain line forward* and the *drain line aft*. The drain line forward

and the drain line aft can be cut off by shutting their respective stop valves located in the pump room.

Proceeding forward from the pump room, we note that the *drain line forward* extends to the forward torpedo room and provides pumping connections for the two bilges and the underwater log well in the after section of the torpedo room. The drain line terminates at the forward bilge manifold with two valves controlling the suction from the poppet valve drain tank and the forward bilge.

The escape trunk drain opens into the forward torpedo room; the water drains directly onto the deck and

The drain water from the gun access trunk, the cable trunk, the periscopes, and the antenna wells empties into the pump room bilge and collects in the sumps from which it is pumped when required.

The drain pump has three points to which it may discharge: 1) overboard discharge, 2) compensating water main, 3) trim system.

In addition, the drain pump is so cross connected with the trim manifold that it can discharge water into the trim system instead of into its own piping. This cross connection permits the use of either the drain pump or the trim pump with either the trim or the drain system, in the event

eventually empties into the bilges.

There are no drain line connections in the forward battery compartment.

The drain line aft extends to the after torpedo room and contains pumping connections to the sumps in the compartments in the after section of the submarine. There are no drain line connections in the after battery compartment. The forward engine room has two bilge sumps connecting with the drain line aft through two individual lines. The after engine room also has two bilge sumps which connect to the drain line by means of two separate lines. In addition to the bilge sump pumping connections, the drain line aft also contains a suction line to the collection tank, making it possible for water from the collecting tank to be pumped out through the drain system.

There is one bilge sump in the motor room.

The drain line aft terminates in the after bilge manifold in the after torpedo room. Here, too, the manifold contains two valves, controlling suction from the forward and after bilge sumps.

Returning now to the pump room, the drain pump suction line carries a branch connection to the pump room bilge manifold. This manifold contains three valves controlling suction front the three pump room bilge sumps.

that one of the two pumps is not in operating condition.

Every branch suction line to the bilge sumps has its own bilge stop valve. When it is desired to pump out certain bilge sumps or wells, the valves leading from them to the drain line and the pump are opened; then the required discharge valves are opened to the overboard discharge, the compensating water main, or the trim system, depending upon the conditions. The drain pump is then started and the pumping begins. After the pumping is completed, the pump is stopped and the valves to the various lines used in the operation are shut.

The drain system can discharge the bilge water directly overboard, into the expansion tank through the compensating water main, or into the trim system through the trim manifold.

Normally, bilge water should not be discharged directly overboard because the oil in it will rise to the surface, indicating the presence of the submarine. Instead, the water should be pumped into the expansion tank, where the water separates from the oil before being discharged overboard.

If the trim system is used to receive the bilge drainage, it is possible to pump this water into the variable ballast tanks. This should not be done normally, because discharging variable tanks to sea during trimming operations will allow bilge oil to rise to the surface, leaving the telltale oil slick.

118

F. DRAIN PUMP

10F1. Source of power. An electric motor, rated at 10 hp and 1,150 rpm, is used to drive the drain pump through a worm and worm gear assembly as shown in Figure 10-5. There are two types of pumps used: one with vertically mounted motor as shown in the large cutaway view of Figure 10-5, and the other with the motor mounted horizontally as shown in the small illustration. The cutaway view shows the mechanical construction of the pump.

10F2. Description. The drain pump is a single-acting duplex reciprocating pump with the cylinders mounted vertically. The two plungers are connected to the crankshaft by connecting rods, so that one plunger completes its downward travel at the moment the other plunger completes its upward travel. As a plunger moves upward in the cylinder, it creates a vacuum (suction) which draws water into the cylinder through the valves from the inlet, or suction, port. When the plunger reaches the top of its stroke and starts its downward travel, the water forces the suction valve down, closing the inlet port, opening the discharge valve, and allowing the water to flow out of the discharge port. At the same time, the second plunger is performing the reverse operation, taking a suction, while the first plunger is discharging. This results in a continuous flow of water through the pump.

An air chamber is provided for each cylinder to smooth out the flow and quiet the pump operation by cushioning the discharge.

Air in the chamber is compressed during the discharge. When the plunger reaches the end of its stroke, expansion of the air tends to keep the water flowing until the reverse stroke begins.

A connection is provided to the 225-pound air system for recharging the chambers. Indicator lights show when the chambers need charging or venting.

10F3. Lubrication. Lubrication of the main and the connecting rod bearings is accomplished by the multiple oiler mounted on the pump casing. Oil is led to the bearings by holes drilled through the crankshaft and connecting rods. The worm gear drive runs in oil, which is cooled by sea water circulating through a coil installed in the worm drive housing.

10F4. Relief valve. The relief valve, set at 225 psi, is mounted on the pump body and protects the pump from excessive pressure in case a valve is shut on the discharge line when the pump is operating.

A drain cock is provided to allow the draining of all water from the pump.

10F5. The drain pump controls. The electrical controls for the drain pump consist of the motor switch, the air chamber pressure indicators, and control panel. All of these are mounted on the port side of the pump room.

The motor switch is equipped with a push button for starting, a push button for stopping, and a signal light that is on when the motor is running. (See Figure 10-6.)

G. VALVES AND FITTINGS

10G1. Drain line stop values. The drain system is provided with two valves, known as the forward and the after drain line stop valves respectively. These valves will put either drain line on service, depending on which section of the boat is to be serviced. The valves are located on the port side of the pump room, forming the connection between the line leading to the suction side of the drain pump and the forward and after drain lines.

The forward drain line stop valve is an angle valve of the disk and seat type with a bolted bonnet, a rising stem, and flanges for connection to the lines. The after drain line stop valve is a globe valve.

10G2. Drain pump overboard discharge valve. When the water collected from the bilges by the drain system is to be discharged directly to the sea, two valves must be opened to provide passage for the drain water.

119

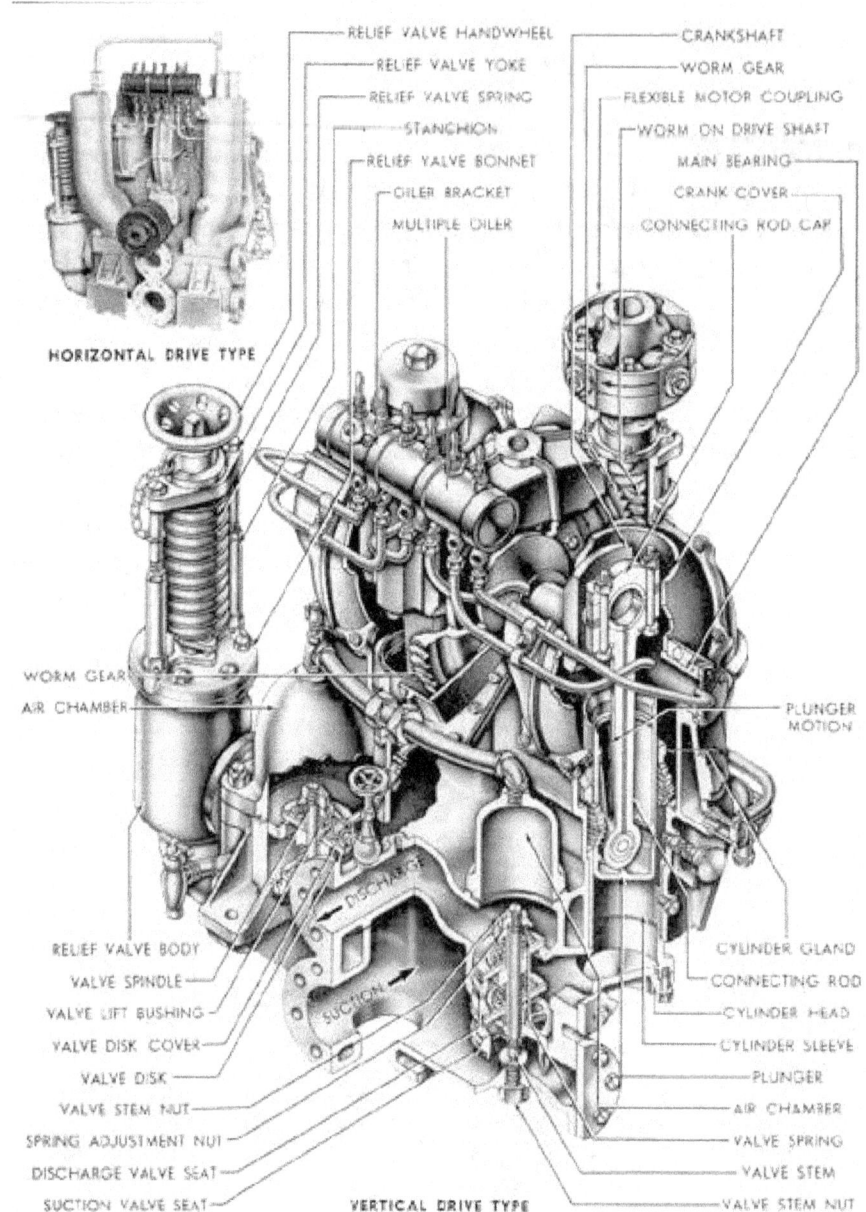

RELIEF VALVE HANDWHEEL
RELIEF VALVE YOKE
RELIEF VALVE SPRING
STANCHION
RELIEF VALVE BONNET
OILER BRACKET
MULTIPLE OILER

CRANKSHAFT
WORM GEAR
FLEXIBLE MOTOR COUPLING
WORM ON DRIVE SHAFT
MAIN BEARING
CRANK COVER
CONNECTING ROD CAP

HORIZONTAL DRIVE TYPE

WORM GEAR
AIR CHAMBER

PLUNGER MOTION

RELIEF VALVE BODY
VALVE SPINDLE
VALVE LIFT BUSHING
VALVE DISK COVER
VALVE DISK
VALVE STEM NUT
SPRING ADJUSTMENT NUT
DISCHARGE VALVE SEAT
SUCTION VALVE SEAT

DISCHARGE

SUCTION

CYLINDER GLAND
CONNECTING ROD
CYLINDER HEAD
CYLINDER SLEEVE
PLUNGER
AIR CHAMBER
VALVE SPRING
VALVE STEM
VALVE STEM NUT

VERTICAL DRIVE TYPE

Figure 10-5. Drain pump.

120

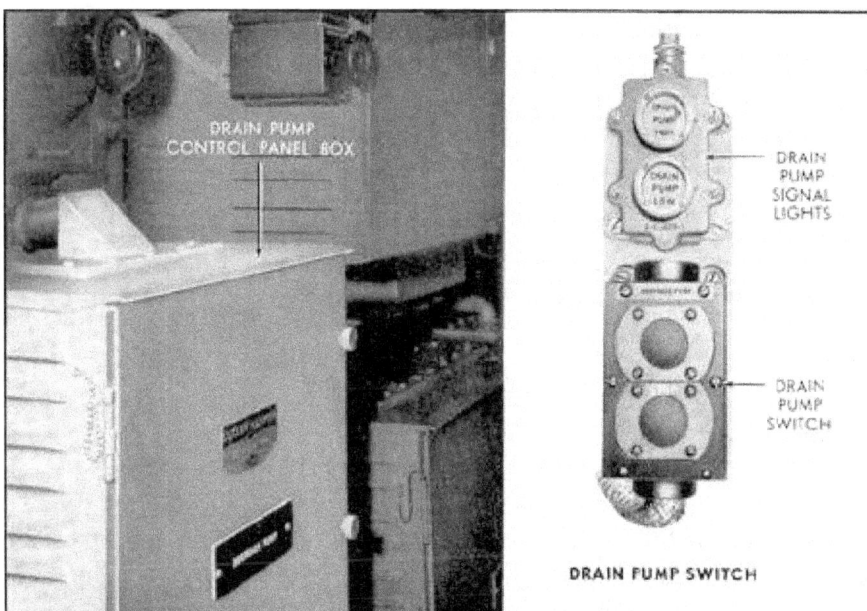

Figure 10-6. Drain pump controls.

The inboard valve is a stop check valve; the second valve is outboard of the first and is known as the drain pump overboard discharge valve. Both are located on the port side forward in the pump room, and are mounted in tandem so that the stop check valve acts as a sea stop for the discharge valve.

10G3. Bilge strainer. Although the purpose of the bilges is to collect excess water, a miscellany of solid material such as flakes of paint and bits of metal inevitably finds its way into the bilges. If this solid matter were to enter the lines of the drain system, it might clog or damage the drain pump. To prevent this, each bilge sump is equipped with a bilge strainer which screens the bilge water before it enters the drain system lines and holds back any large particles.

10G4. Macomb strainer. Although the bilge strainers discussed in Section 10-G3 will prevent pieces of solid material larger than 1/2 inch from

might clog or damage the drain pump. Such material is filtered out of the drain system by the Macomb strainers.

10G5. Drain line sight glass. The drain sight glass provides a means of visually determining the amount of oil of solid matter in the bilge water as it flows through the lines of the drain system. It consists of a cross-shaped casting, two ends of which are flanged and connected to the drain lines. The other arms are fitted with glass plates forming a window to allow inspection of the water in the drain line.

10G6. Underwater log well suction line and sump. The water that collects in the underwater log well is pumped out by the underwater log well suction line. This line extends from the drain line and runs athwartships along the after bulkhead of the forward torpedo room to the underwater log well.

It is equipped with a bilge strainer which is fitted into the well. A stop check valve is mounted in the line

entering the drain system, it is necessary to screen the water again to remove any smaller particles of debris that between the well and the forward drain line which is opened to pump the underwater log well.

121

11
AIR SYSTEMS

A. GENERAL DATA

11A1. Importance of the air systems to submarines. The importance of the air systems to a submarine cannot be overemphasized, for virtually every function in the diving and surfacing procedure stems initially from air provided by one or more of the air system; to cite a few:

a. The main hydraulic system operates because of the air pressure maintained in the air-accumulator flask.

b. Torpedoes are discharged from the submarine by air.

c. Tanks are blown by air.

d. The main propulsion engines are started by air.

Air, or more specifically, compressed air, is necessary to surface, submerge, attack, and cruise. In addition, compressed air together with oxygen is used to revitalize the air in the ship after long periods of submergence. *Pressure in the boat*, a test for tightness, utilizes air.

The air systems represent, therefore, the most versatile of all systems aboard a more operations than any other single system.

11A2. Basic principles of compressed air. The actuating force of the air systems is compressed air, which, as the name implies, is air under pressure confined within the limits of a container. The force required for compression of the air is provided by the high-pressure air compressors, a simple machine that compresses air by means of a series of pistons designed so that one or more pistons discharges air into another for further compression and finally through lines to banks for storage. Air can be compressed easily aboard a submarine, as it requires a relatively small plant and comparatively simple equipment. It can be stored at any convenient place and is always ready for use. Its action can be regulated to produce a low or high-pressure, and yet it has enough elasticity or compressibility to cushion its impact against the equipment it operates. It consumes no valuable materials and can be supplied to any part of the submarine simply by extending a line from the air supply. Air, once stored, requires no further expenditure of energy for operation, but rather is a source of power to other equipment.

submarine, in that they are capable of performing, either as primary or secondary functions,

B. TYPES AND RELATIONSHIPS OF AIR SYSTEMS

11B1. General information. There are five air systems on the submarine: the 3,000 pound high-pressure and torpedo impulse system; the 600-pound main ballast tank (MBT) blowing system; the 225-pound service air system (ship's service air); the 10-pound main ballast tank (MBT) blowing system; and finally, the salvage air system. (See Figure A-13.)

The 600-pound MBT blowing system and the 225-pound service air system receive their supply of air from the 3,000-pound air system. The 10-pound MBT blowing system is an independent system with its own low-pressure blower. The internal compartment

salvage air system is dependent upon the 225-pound service air system, while the external compartment salvage air system is entirely dependent upon an outside source for its supply of air.

11B2. The 3,000-pound and torpedo tube impulse air system. This system consists of the 3,000-pound high-pressure compressors, the high-pressure manifold, the interconnecting piping, valves, and compressed air banks. The main function of the 3,000-pound air system is to compress, store and supply air at the maximum pressure of 3,000 pounds per square inch for use within the 3,000-pound, the 600-pound, and the 225-pound systems.

The 3,000-pound air system also supplies air to the hydraulic accumulator air loading manifold and to the forward and after 600 pound Grove reducing valves which supply the forward and after torpedo tube impulse charging manifolds.

The 3,000-pound air system is equipped with air external charging connection so that the system may be supplied with air from an outside source.

11B3. The 600-pound MBT blowing system. The only function of the 600-pound MBT blowing manifold and system is to remove water ballast front the main ballast tanks or the fuel ballast

aboard the submarine. The 225-pound system consists of the 225-pound service air manifold, interconnecting piping, and various valves.

11B5. The 10-pound MBT blowing system. When the submarine has surfaced, the 10-pound main ballast tank blowing system is used to conserve the compressed air stored in the ship's air banks. The system consists of its own low-pressure blower, control manifold, and piping to the various main ballast tanks. The 10-pound system is operated only after the submarine has, surfaced sufficiently to permit the opening of induction valves and hatches.

11B6. The salvage air system. This

tanks when used as main ballast tanks. It receives its supply of compressed air from the high-pressure system through the distributing manifold.

11B4. The 225-pound service air system. The 225-pound service air system or, as it is sometimes called, ship's service air, is so called because, in addition to blowing the variable group of tanks, it provides the compressed air for all the miscellaneous services

system is actually three separate systems: the MBT external salvage, the compartment external salvage, and compartment internal salvage. The external salvage connections permit compressed air from an outside source to be supplied to the tanks and/or compartments, while the internal salvage system utilizes the ship's air for, compartment salvage only.

C. HIGH-PRESSURE AIR AND TORPEDO IMPULSE AIR SYSTEMS

11C1. General description. Figure A-14 shows the location and relationship of the individual units that comprise the high-pressure 3,000-pound air system. It should be noted that 3,000 pounds is the maximum working pressure of the system end not a constant pressure. Actually, the pressure may vary between 1,500 and 3,000 psi. The system is hydrostatically tested to 4,500 psi or 150 percent of the working pressure. The system extends from the high-pressure air compressors in the pump room to the receiving and distributing manifolds in the control room, and from there forward to the torpedo impulse air system in the forward torpedo room, athwartship to the air banks, and aft to the after torpedo room.

In Sections 11C1 through 11C4, a more detailed description is given of the general layout of the high-pressure air system. The control room, the air banks, and the torpedo impulse air system fore and aft of the control room are discussed.

11C2. Manifolds and lines. The high-pressure manifold, made up of a receiving manifold and two distributing manifolds, is mounted on the starboard side of the control room. The receiving manifold receives air up to 3,000 psi from two high-pressure air compressors, and directs it to the air banks where it is stored. The capacity of each compressor is 20 cubic feet per hour at 3,000 psi. As the air is needed, it flows back through the same piping to the receiving manifold where it is directed to the distributing manifold. This operation is controlled by the valves on the manifold.

The 3,000-pound service air lines supply air at a pressure up to 3,000 psi to the forward and after torpedo rooms, to the engine starting flasks, and to the reducing valves in each engine room which furnish 500-pound air used in starting the diesel engines.

The distributing manifolds distribute air to the safety and negative tank blow lines, the main ballast tanks blow manifold, the

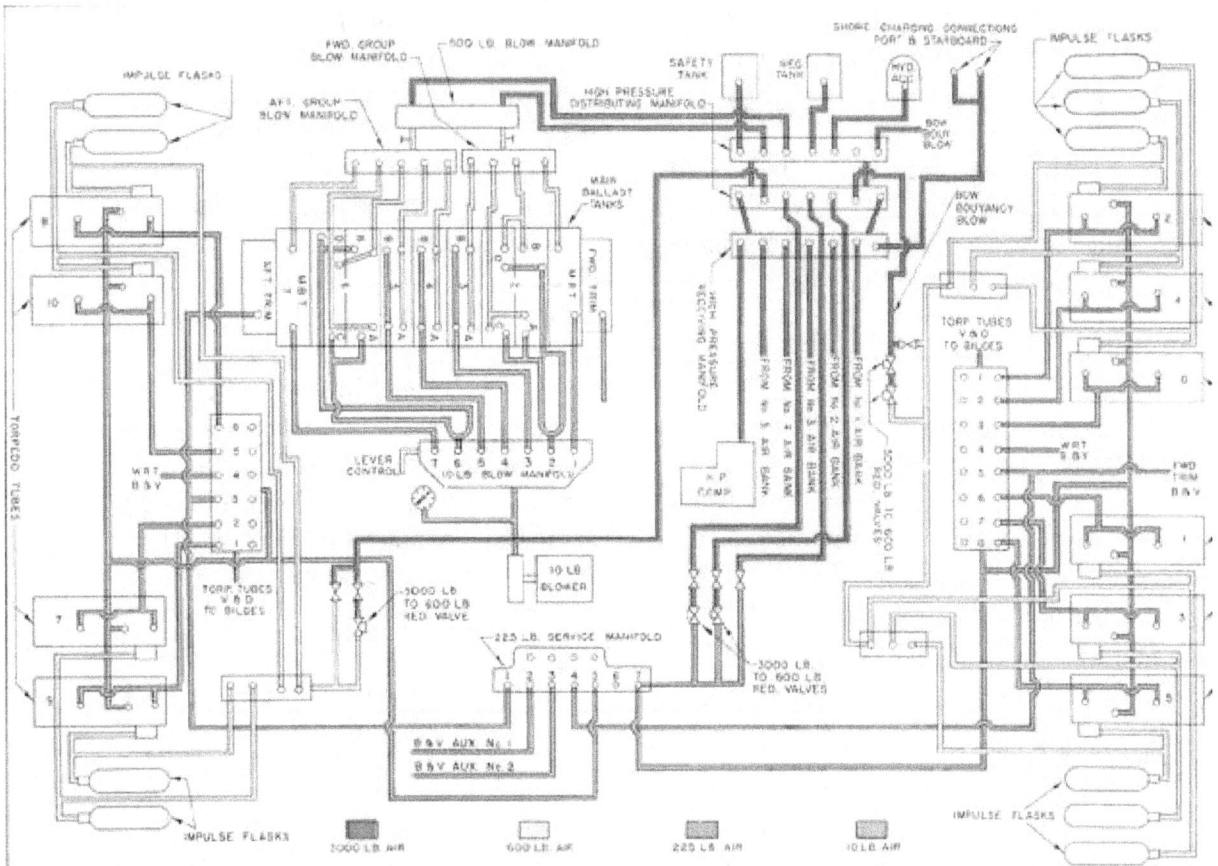

Figure 11-1. Compressed air systems.

124

hydraulic accumulator air flask, the high-pressure air bleeder, the bow buoyancy tank blow line, the 225-pound service air system, and the forward and after 3,000-pound service air lines.

11C3. Air banks. Each of the five air banks consists of seven flasks, with the exception of the No. 1 air bank which has eight. Each flask is provided with a drain valve. The total capacity of the air banks is 560 cubic feet.

The No. 1 air bank is located inside

charging manifolds and six impulse flasks connected by lines to the manifolds. The impulse flasks are mounted above the pressure hull in the superstructure forward. One impulse flask charging manifold is located on the port side of the torpedo room and the other on the starboard side. Each manifold is used to charge three flasks with 600-pound air.

The 3,000-pound air service line aft, extending from the distributing manifold, ends with a 3,000-pound to 600-pound reducing valve, through which air is furnished to the after the pressure hull, with four flasks in each battery, compartment. The other four air banks are located in the main ballast tanks. (See Figure A-14.)

11C4. Torpedo impulse air system. The torpedo impulse air system stores and controls the air used to discharge the torpedoes from the tubes.

The 3,000-pound air service line forward, extending from the distributing manifold, ends with a 3,000-pound to 600-pound reducing valve, from which a line leads to the forward torpedo impulse air system. This system is composed of two impulse flask

torpedo impulse air system. This system consists of one impulse flask charging manifold with lines leading to the four impulse flasks provided for the four after torpedo tubes. The impulse flasks are mounted below the after torpedo room deck; the manifold is located on the starboard side.

In both the forward and the after sections of the torpedo impulse system, a bypass valve and line are provided, leading from the 3,000-pound air service line to the charging manifold. The bypass valve and line allow the charging of the impulse flasks in the event of failure of the reducing valves.

D. THE 600-POUND MAIN BALLAST TANK BLOWING SYSTEM

11D1. General description. The main ballast tanks are filled with sea water when the submarine is submerged. These tanks cannot be pumped. Therefore, when the submarine is surfacing, compressed air must be used to blow the water out through the flood ports to the sea.

Two separate systems are provided to blow the main ballast tanks. This section describes the first of these, the 600-pound MBT (main ballast tank) blowing system. The second system, the 10-pound MBT blowing system, is used only when the ship is surfaced.

Figure A-15 shows the location of the lines and component parts of the 600-pound MBT blowing system. The system is inside the pressure hull and extends from the MBT blowing manifold in the control room fore and aft along the starboard side to the main

distribution control unit of the system. It is located on the starboard side of the control room with its pressure gage next to it. The piping mounted directly above the manifold connects the MBT blowing manifold with the high-pressure air manifold through two hammer valves.

The maximum working pressure of the 600-pound main ballast tank blowing system is 600 psi. It is tested hydrostatically to a pressure of 1,000 psi, or 166 percent of the maximum working pressure.

Air at bank pressure (1,500 to 3,000 psi) passed through two manually operated hammer valves and two group stop check valves to the 600-pound MBT blowing manifold. The flow of the air is regulated by the hammer valves so that it is delivered at the required pressure. Normally, only one hammer valve is used for blowing; in case one does not

ballast tanks and fuel ballast tanks.

The MBT blowing manifold is the

supply enough air or in case of failure, the other hammer valve can be used. The

group stop check valves permit the blowing of tanks by groups. The manifold is protected by a sentinel valve and two relief valves set to blow when the pressure in the 600-pound system reaches 750 psi. A sentinel valve is set to blow at a pressure of 610 psi. When the sentinel valve opens, it acts as a relief valve for comparatively small rises in pressure and gives notice of excessive pressure in the system.

To supply air to the 600-pound MBT blowing system, one of the hammer valves is opened. The valve permits air from the 3,000 pound manifold to enter the MBT blow manifold at a reduced pressure. The pressure gage of the MBT blow manifold is closely watched to guard against the pressure exceeding 600 pounds.

The depth at which the submarine is operating will have a direct effect on the resistance offered to the air in blowing the main ballast tanks and, therefore, the pressure will be built up within the system more rapidly at greater depths than it will on the, surface. Since the hammer valve regulates the volume of the air entering the 600-pound MBT blowing system, while the resistance offered to this air varies with submerged depth, it follows that, when submerged at great depths, the hammer valve must be opened cautiously, otherwise the pressure within the system will build up rapidly and exceed the safe

maintain the required pressure. When blowing is finished, the hammer valve is shut.

Blow lines extend from the forward section of the 600-pound MBT blow manifold to tanks No. 1 MBT, Nos. 2B and 2D MBT. Nos. 2A and 2C MBT, and Nos. 3A and 3B FBT. From the after section of the manifold, blow lines extend to tanks Nos. 4A and 4B FBT, Nos. 5A and 5B FBT, Nos. 6B and 6D MBT, Nos. 6A and 6C MBT, and No. 7 MBT. Any tank or any combination of tanks can be blown by opening the required individual tank valve, or valves, the group valves, and finally the hammer valves.

When the submarine is rigged for diving, all the blow valves on the manifold, except the fuel ballast tanks valves, are open, as are the two group stop check valves. The individual regulator valves at the main ballast tanks are open, while the MBT blow hammer valves on the 600-pound manifold are shut. The two supply valves to the 600-pound MBT blow manifold on the distributing manifold are open.

To operate the 600-pound MBT blow system, the hammer valve is opened and air is admitted to the blow manifold, from which it is directed to the main ballast tanks by the lines of the system.

Each blow line is provided with a regulator valve at the point where it enters the tank. The regulator valve acts

working pressure. When the gage indicates that the pressure is dropping, the hammer valve is opened wider to as a combination stop and check valve and is equipped for securing the stop in any position required to equalize the flow of air into the tanks.

E. THE 225-POUND AIR SYSTEM (SHIP'S SERVICE AIR)

11E1. General description. The 225-pound service air system, known as the *ship's service air system*, performs or controls many operations other than those discussed in the sections dealing with the 3,000-pound, the torpedo impulse, and the 600-pound air systems. The 225-pound air system provides the air for approximately 100 operations. The system extends from the forward torpedo room to the after torpedo room, with service connections in every compartment of the vessel, and supplies air at pressure ranging. from 225 psi to 8 psi. The center of direction of the system, the 225-pound service air manifold, is located in the control room. The 225-pound system is hydrostatically tested to 350 psi, or 155 percent of its maximum working pressure of 225 psi.

FigureA-16 shows the, location and relationship of the parts of the system, as well as their nomenclature.

The discussion of the ship's service air system starts with the control room, describing each component part of the system located there, and explaining its function in the operation of the submarine. A similar

procedure is followed for each of the compartments of the vessel, proceeding first forward and then aft of the control room.

11E2. Control room. The 225-pound service air manifold is located in the control room on the starboard side aft of the high-pressure manifold. This manifold receives its supply of air through two Grove pressure-reducing valves which reduce the high-pressure air from 3,000 psi to 225 psi. Stop valves are provided on the low-pressure side of the 225-pound Grove reducers, cutting them off the 225-pound system. This permits removal of a Grove reducer without impairing the operation of the 225-pound system.

supplies 225-psi and air to the hydraulic oil supply volume tank, the signal ejector, the drain pump air domes, the negative tank blow valve, and the sea pressure and depth gage blows.

The other branch of the line from the Grove reducing valves supplies air to the 225-pound service air manifold. This air is directed by means of valves to the forward and after service air mains, the auxiliary tank blow and vent lines, and the forward and after trim tank blow and vent lines. A hose connection to the manifold provides for air supply from the shore or tender.

A reducing valve from the forward service air main furnishes air at pressure

The 225-pound service air manifold can also be supplied from the 225-pound bypass which is controlled by a manually operated 225-pound bypass valve located at the high-pressure distribution manifold. Where the 225-pound bypass is used, the high-pressure air bypasses the reducing valves and is admitted directly into the 225-pound system. The bypass valve is only partly opened so that the pressure can be built up gradually. The 225-pound manifold pressure gage must be constantly watched and the pressure must never be allowed to go beyond 225 psi.

The 225-pound service air system is protected by one sentinel valve and two relief valves located on the line between the Grove reducers and the 225-pound manifold.

When the air within the 225-pound system reaches a pressure of approximately 250 psi, the sentinel valve opens, allowing the excess air to escape into the compartment. The sentinel valve has a comparatively small capacity and serves primarily to warn that the normal working pressure is exceeded. If the rise in pressure is rapid and above the capacity of the sentinel valve, the two relief valves, set to operate at 275 psi, open and allow the excess air to escape into the compartment.

The relief valves and the sentinel valve shut automatically when the normal working pressure is restored.

The supply line from the Grove reducing valves has two branches. One branch

of 100 psi to a connection for pneumatic tools. A bypass is provided for emergency operations, with a relief valve set to open at a pressure of 110 psi as a protection against excessive pressure. It also carries a connection supplying air to the whistle and siren.

The after service main has branch connections to the sea pressure gage and to the compartment air salvage. It also supplies air through a reducing valve at a pressure of 12 psi to fresh water tanks No. 3 and No. 4. A bypass is provided for emergency operation, with a relief valve set to open at a pressure of 15 psi.

11E3. Forward battery compartment. In the officers' quarters, the forward service air main supplies the compartment air salvage valve mounted on the after bulkhead. This valve can be operated from either side of the bulkhead. A branch of the service line, passing through an 8-pound reducing valve, supplies air at a pressure of 8 psi to the four battery water tanks Nos. 1, 2, 3, and 4 in the forward battery compartment. A bypass line is provided for emergency operation, with a relief valve set to open at a pressure of 10 psi.

11E4. Forward torpedo room. In the forward torpedo room, the forward service air main extends to the torpedo tube blow and vent manifold. The service main is also provided with branch lines to the torpedo stop cylinders, the escape trunk blow, the volume tank, the sanitary tank, the QC and JK sea chests, the underwater log, the compartment air salvage valve, and the fuel oil blow and vent manifold. Two other branch lines, equipped with reducing valves and bypass lines, furnish air to the pneumatic tool connection at 100 psi, and to the No. 1 and No. 2 fresh water tanks at 12 psi. The line to the escape trunk supplies air for the ship's diver's air, connection, and a blow and vent line for the escape chamber. The forward trim tank blow and vent line from the 225-pound manifold in the control room terminates at the forward trim tank and connects with the forward trim tank blow and vent line from the forward torpedo tube blow and vent manifold.

11E5. After battery compartment. The galley and mess room compartment has one connection from the after service main which supplies air to the blow and vent manifold for fuel ballast tanks 3A, 3B, 4A, and 4B. A second connection enters an 8-pound reducing valve and supplies air at 8 psi to the four battery fresh water tanks Nos. 5, 6, 7, and 8 located in the after battery compartment. Bypass is provided for emergency use with a relief valve set to open at 10 psi. The lines for blowing and venting the auxiliary ballast tanks connect from the 225-pound manifold to the auxiliary ballast tank and stop valves located at the tank top in this compartment.

11E6. Crew's quarters. In the crew's quarters, the after service main supplies air to the crew's forward

with a relief valve set to open at 15 psi. In addition, the forward engine room is provided with a pneumatic tool connection equipped with a 100-pound reducing valve and a bypass for emergency. A relief valve, set to open at 110 pounds, safeguards the line against excessive pressure.

11E8. After engine room. In the after engine room, the after service main has direct connections to the compartment air salvage valve, the auxiliary engine shutdown, and the air manifold which controls the blowing and venting of the Nos. 6 and 7 normal fuel oil tanks, the expansion and the collecting tanks. A relief valve, set to open at 15 psi, protects No. 6 and No. 7 normal fuel oil tanks and the collecting and expansion tanks against excessive internal pressure. A pneumatic tool connection is also provided, equipped with a 100-pound reducing valve, 110-pound relief valve, and a bypass line, to supply air at 100 psi.

11E9. Maneuvering room. The maneuvering room contains lines extending from the after service main to the after water closet, the compartment air salvage valve, and the main engine shutdown connection.

11E10. After torpedo room. In the after torpedo room, the service air main has branches leading to the 225-pound compartment air supply valve for escape hatch, the torpedo tube stop cylinders, the volume tank, and the pneumatic tool connection. The pneumatic tool connection is provided with a 100-pound reducing valve and a bypass protected by a 110-pound relief valve.

water closet and the No. 2 sanitary tank blow line. The sanitary tank is equipped with a relief valve set to open at 105 psi.

11E7. Forward engine room. The forward engine room has direct connecting lines from the after service main to the compartment air salvage valve, the No. 5A and No. 5B fuel ballast tank manifold, the exhaust valve operating gear, and the lubricating oil tanks blow and vent manifold. The supply to the fuel oil manifold is protected by a relief valve set to open when the pressure exceeds 15 psi. The air for the lubricating oil manifold is reduced to 13 psi by a reducing valve. A bypass is provided for emergency operation,

The service air lines terminate at the after torpedo tube blow and vent manifold.

The after trim tank blow and vent line, which extends from the 225-pound manifold in the control room, connects with the after trim tank by a branch line extending to the torpedo tube blow and vent manifold, similar to that of the forward torpedo room.

The compartment air salvage valves are so mounted on the transverse bulkheads of each compartment that they may be operated from either side, releasing air into the compartment from which they are worked, or into the adjoining compartment. The compartment air pressure gages are also mounted

on either side of the bulkheads to permit a reading of air pressure in the adjoining compartment.

All manifolds and lines equipped with reducing valves and blow valves are provided

with pressure gages. All fuel oil, lubricating oil, collecting, expansion, sanitary, and variable tanks are provided with pressure gages located in the various rooms and compartments.

F. THE 10-POUND MAIN BALLAST TANK BLOWING SYSTEM

11F1. General description. The 10-pound MBT blowing system is used to remove water from the main ballast tanks when the submarine is on the surface. It completes the work started by the 600-pound MBT blowing system, thus saving high-pressure air.

The 10-pound MBT system (Figure A-17) consists of a low-pressure blower located in the pump room, a manifold, and blow lines to each of the tanks

The list control dampers are used to correct a list during blowing of the main ballast tanks. The list control dampers adjust the amount of air admitted into the port or starboard ballast tanks of the No. 2 and No. 6 MBT group, increasing or decreasing the rate at which the tank is blown. The dampers are located at the Y outlet connections on the 10-pound blow manifold.

Both list dampers are attached to a shaft

serviced by the system. The low-pressure blower furnishes compressed air to the manifold in the control room at a pressure of approximately 10 psi. The manifold distributes the air supplied by the blower to the ballast tanks through nine pipe lines which pass through the hull directly above the manifold and extend outside the pressure hull under the superstructure deck.

The air supply to the manifold is controlled by the flapper valve. The manifold and the valves are designed to withstand sea pressure if any of the blow lines fail.

The nine low-pressure lines have lever-operated flapper valves (10-pound blow valves) at the point where they pass through the hull, and swing check valves where they join the main ballast tank (MBT) vent risers.

Gate valves, controlled from the superstructure deck, are installed in the lines leading to main ballast tanks 2A, 2B, 2C, 2D, 6A, 6B, 6C, and 6D.

which runs through the manifold chamber. The shaft is operated by a hand lever at the after end of the manifold. The handle assembly consists of a push rod at the top of the handle, a handle, a spring, a latch, a name plate, and a bracket. Pressing down the push rod releases the spring, lifting the latch, and leaving the lever free to move inboard or outboard. As the shaft turns, the list dampers are swung to shut one port or open both ports of the Y.

The movement of the lever and the attached connecting rod turns the shaft by means of an offset arm. Outboard movement of the lever causes the damper to restrict the flow of air to the starboard side. Inboard movement of the lever causes the damper to restrict the flow of air to the port side. The normal position of the damper is neutral, allowing equal flow to both sides.

G. SALVAGE AIR SYSTEM

11G1. General description. The submarine is provided with a salvage air system for use, in salvage operations.

The salvage air arrangements provide external salvage facilities for use by outside salvage agencies (divers, and so forth) and also internal facilities for use by the crew of the submarine or by a diver, after he succeeds in entering the vessel.

Figure A-18 shows in schematic form the location and relation of the component parts that comprise the salvage air system.

Two external high-pressure air connections, located on each side of the conning tower, provide a means of supplying high-pressure air from the salvage ship to the high-pressure (3,000-pound) receiving manifold. This air can be directed by personnel

COMPARTMENT SALVAGE DECK PLATE MARKINGS *

Compartment	Number of Screw Heads	
	HIGH CONNECTION	LOW CONNECTION
Officers' quarters	1	2
Forward torpedo and control room	3	4
Crew's quarters	5	6
Forward engine room	7	8
After engine room	9	10
Maneuvering room	11	12
After torpedo room	13	14

TANK SALVAGE DECK PLATE MARKINGS

Tank	Number of Lugs
MBT 2A and 2B	1
MBT 1	2
MBT 2C and 2D	2
MBT 6A and 6B	3
MBT 6C and 6D	4
MBT 7	5

10-POUND DECK PLATE BLOW MARKING

Tank	Number of Lugs
MBT 2A and 2B	2
MBT 2C and 2D	4
MBT 6A and 6B	6
MBT 6C and 6D	8

* The markings used for a particular submarine may be obtained from the vessel's air salvage systems plans.

inside the vessel to the 600-pound blow manifold for use in blowing the main ballast tanks. and to the 225-pound service air manifold for use in outside or by a handwheel from within the compartment. In salvaging operations, air hoses can be attached to the valve fittings to supply the ship with blowing water from flooded compartments by means of the compartment salvage air valves.

Each main ballast tank has an external salvage valve with a blow line connection extending up to a plate set in the deck. In salvaging, air hose lines from the salvage ship are attached to the pipe fitting and the valve is opened, thus enabling the rescue vessel to blow the ballast tanks free of water.

Each compartment of the submarine has two external compartment salvage valves, one at either end of the compartment. A salvage line from each valve extends through the hull to a deck plate where it is provided with a capped male fitting, similar to those of the main ballast tank salvage lines. The valve can be operated by a socket wrench from the

air for breathing, pumping, or circulating purposes.

Compartment salvage air valves are located on each bulkhead between compartments, for use in blowing individual compartments. The 225-pound air is supplied to these compartment salvage air valves by lines extending from the forward and after service air lines. The arrangement of the valves permits the release of air from either side of the bulkhead into the adjacent compartment. Pressure gages are installed on both sides of the bulkhead near this valve arrangement to indicate the pressure in the adjoining compartment.

All external salvage valve deck plates are identified by lettering and round screw heads, and special lugs cast on the plates for touch identification.

131

12
MAIN HYDRAULIC SYSTEM

A. INTRODUCTION

12A1. The increasing use of hydraulic power in the modern submarine. In the development of the submarine from pre-war classes, many changes and improvements have occurred. One of the outstanding differences is the large variety of submarine devices that are now operated by hydraulic power. In early classes, there was no hydraulic system, power requirements were met by means of air or electricity. Along with the

stopping of the driving mechanism is demanded, electric motors have a tendency to overtravel, or drift, making fine control difficult to achieve. A further disadvantage in the operation of electrical units is the noise made in starting and stopping by relays and magnetic brakes, and by shafting and other mechanical power transmission units.

c. *Pneumatic power.* Since compressed

steady improvement in submarine design has gone a constant extension and diversification of the use of hydraulic power.

12A2. Other sources of power available on submarines. What is the reason for this noticeable trend toward hydraulics? Obviously, hydraulic actuation is not the only means of transmitting power throughout the submarine. The tasks now being done by the hydraulic system were originally performed by hand, electricity, or compressed air.

a. *Hand power.* Some equipment on a submarine is still operated exclusively by hand, but this practice is rapidly disappearing. This is because the power requirements exceed that which manual effort can provide for long periods of time, and because power operation is faster and can be remotely controlled, greatly reducing the necessary communication between crew members.

b. *Electric Power* Since the electrical plant is such an important part of the submarine power system and must be used for propulsion in any event, it would be reasonable to expect that electricity would also be used to operate all of the auxiliary equipment as well.

Electricity is ideally adapted for submarine equipment having few or no moving parts; that is, lamps, radios, cooking facilities, and similar devices. But, it is not so ideal when it is necessary to move heavy apparatus such as the rudder and bow and stern planes, because heavy bulky electrical units are required. Also, when

air must also be used aboard a submarine for certain functions, this system, comprising the compressors, high-pressure air flasks, and air lines, offers another source of auxiliary power. However, pneumatic or compressed air power also has definite shortcomings. Pressure drop caused by leakage, and the mere fact that air is a compressible substance, may result in *sponginess* or lag in operation. The high pressure necessary for compressed air storage increases the hazard of ruptured lines, with consequent danger to personnel and equipment. Another disadvantage of air systems is that the air compressors require greater maintenance than others and are relatively inefficient.

d. *Comparative advantages of hydraulic power.* Hydraulic systems possess numerous advantages over other systems of power operation. They are light in weight and are simple and extremely reliable, requiring a minimum of attention and maintenance. Hydraulic controls are sensitive and afford precise controllability. Because of the low inertia of moving parts, they start and stop in complete obedience to the desires of the operator, and their operation is positive. Hydraulic systems are self-lubricated; consequently, there is little wear or corrosion. Their operation is not likely to be interrupted by salt spray or water. Finally, hydraulic units are relatively quiet in operation, an important consideration when detection by the enemy must be avoided.

Therefore, in spite of the presence of the two power sources just described, hydraulics makes its appearance on the submarine

instantaneous

because its operational advantages, when weighed against the disadvantages listed for electricity and air, fully justify the addition of this third source of power in the submarine.

12A3. Hydraulic fluids. Almost any free flowing liquid is suitable as a hydraulic fluid, if it does not chemically injure the hydraulic equipment. For example, an acid, though free flowing would obviously be unsuitable because of corrosion to the metallic parts of the system.

Water as a possible hydraulic fluid, except for its universal availability, suffers from a number of serious disadvantages. One is that it freezes at a relatively high temperature, and in freezing expands with tremendous force, destroying pipes and other equipment. Also, it rusts steel parts and is rather heavy, creating a considerable amount of inertia in a system of any size.

The hydraulic fluid used in submarine hydraulic systems is a light, fast-flowing lubricating oil, which does not freeze or lose its fluidity to any marked degree even at low temperatures, and which possesses the additional advantage of lubricating the internal moving parts of the hydraulic units through which it circulates.

12A4. Basic units of a hydraulic system. A simple, hydraulic system will necessarily include the following basic equipment, which, in one form or another, will be found in every

5. A check valve placed in the line to allow fluid motion in only one direction.
6. *Hydraulic lines*, that is, piping to connect the units to each other.

While the functions performed by these six units are typical of every hydraulic system, the units are not always identified by similar names, but rather by names descriptive of the specific operation they perform. The submarine hydraulic system is really four distinct systems: the main hydraulic system, the bow plane tilting system, the stern plane tilting system, and the steering system.

The *main hydraulic system* performs the bulk of the hydraulic tasks aboard a submarine. Lines from the central power source radiate through the ship to convey fluid under pressure for the operation of a large variety of services. The vent valves of the main ballast, fuel oil ballast, bow buoyancy and safety tanks, and the flood valves of the negative and safety tanks are hydraulically opened and shut by power from the main system. It also operates the engine induction and ship's supply outboard valve, the outer doors of the torpedo tubes, the bow plane rigging gear the windlass and forward capstan, the raising and lowering of the echo ranging and detecting apparatus (sound heads), and the main engine exhaust valves on earlier classes of submarines. In the latest installations, the main engine exhaust valves are operated by pneumatic-hydraulic or air-cushion units. In an emergency, the main hydraulic system is also used to supply power for the steering system and for the tilting of

hydraulic system.

1. A reservoir or supply tank containing oil which it supplies to the system as needed, and into which the oil from the return line flows.
2. A pump which supplies the necessary working pressure.
3. A hydraulic cylinder or *actuating cylinder* which translates the hydraulic power developed in the pump into mechanical energy.
4. A Control valve by means of which the pressure in the actuating cylinder may be maintained or released as desired.

the bow and stern diving planes, although these systems normally have their own independent power supply units.

On the latest classes, the periscopes and antenna masts are also hydraulically operated as units of the main hydraulic system. (In earlier classes, they are electrically operated.)

To perform these numerous tasks, a variety of valves, actuating cylinders, tanks, and manifolds are required, as well as the pumps for building up the required power. The units in the main hydraulic system fall conveniently into five groups:

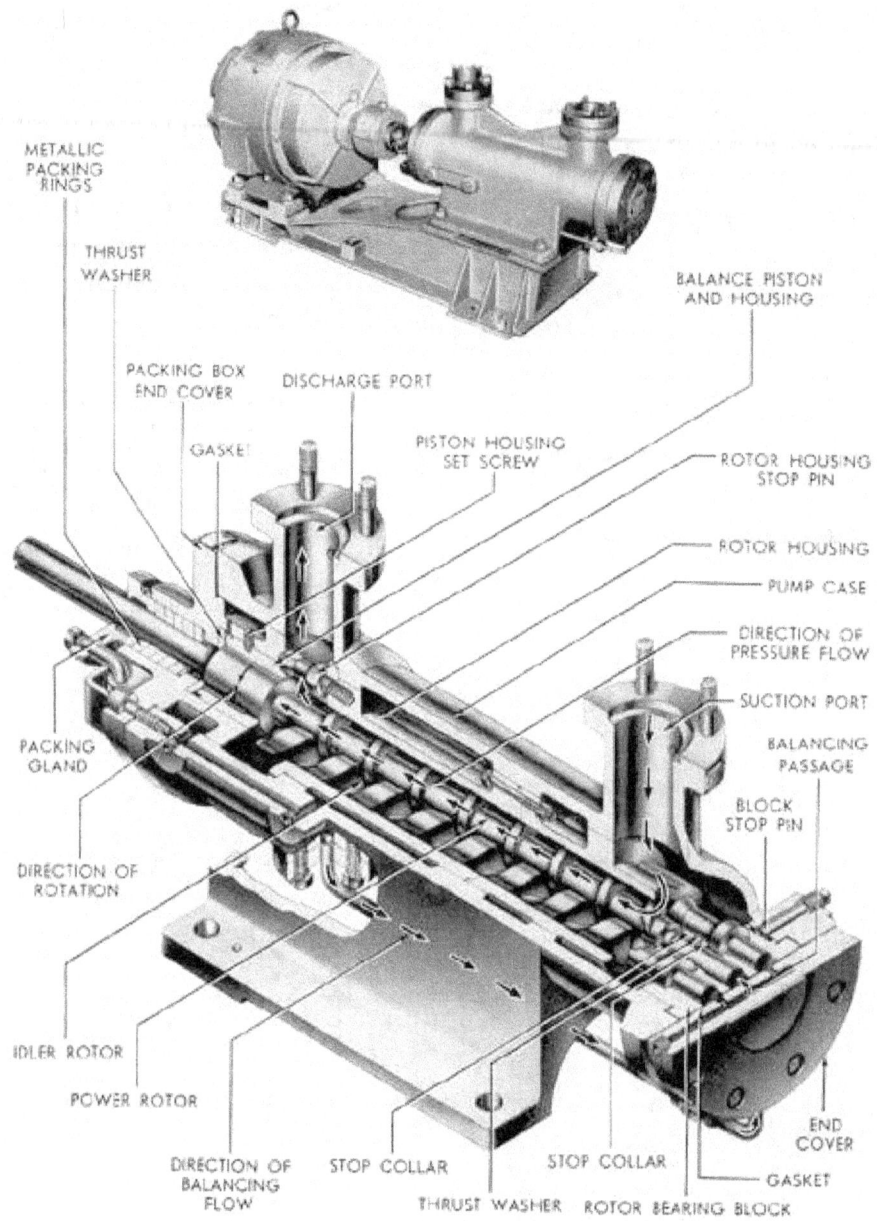

METALLIC
PACKING
RINGS

THRUST
WASHER

PACKING BOX
END COVER

GASKET

DISCHARGE PORT

PISTON HOUSING
SET SCREW

BALANCE PISTON
AND HOUSING

ROTOR HOUSING
STOP PIN

ROTOR HOUSING

PUMP CASE

DIRECTION OF
PRESSURE FLOW

SUCTION PORT

BALANCING
PASSAGE

BLOCK
STOP PIN

PACKING
GLAND

DIRECTION OF
ROTATION

IDLER ROTOR

POWER ROTOR

DIRECTION OF
BALANCING
FLOW

STOP COLLAR

STOP COLLAR

THRUST WASHER   ROTOR BEARING BLOCK

GASKET

END
COVER

Figure 12-1. IMO pump.

134

---

1. Power generating system.

2. Floods and vents.

3. Periscope and radio mast hoists.

4. Forward and after service lines.

5. Emergency systems.

Figure<u>A-19</u> shows a schematic view of the main hydraulic system in the submarine.

B. POWER GENERATING SYSTEM

12B1. General arrangement. The

i. The accumulator air flask, located in

power generating system comprises a group of units, the coordinated action of which provides the hydraulic power necessary for the operation of the main hydraulic system. It consists of the following principal parts:

a. The IMO pumps, located in the pump room, which supply hydraulic power to the system.

b. The main supply tank, located in the control room, which contains the oil needed to keep the system filled.

c. The accumulator, located in the pump room, which accumulates the oil from the pump and creates pressure oil which is maintained at a static head for instant use anywhere in the system.

d. The main hydraulic manifolds, located in the control room, which act as distribution and receiving points far the oil used throughout the system.

e. The pilot valve, a two-port, fitted lap-fitted trunk, cam-operated slide valve, located in the pump room, which directs the flow of oil that causes the automatic bypass valve to open or shut.

f. The automatic bypass and nonreturn valves which are located in the pump room. The automatic bypass valve directs the flow of pressure oil in obedience to the action of the pilot valve. The nonreturn valve prevents the oil from escaping through the open automatic bypass.

g. Cutout valves, serving various purposes throughout the system and nonreturn valves which allow one-way flow.

the pump room, which serves as a volume tank for the accumulator, allowing the air to pass to and from it when the accumulator is loading or unloading.

12B2. IMO pump. Hydraulic systems need, in practice, some device to deliver, over a period of time, and as long as required, a definite volume of fluid at the required pressure.

The IMO pump (Figure 12-1) is a power-driven rotary pump, consisting essentially of a cylindrical casing, horizontally mounted, and containing three threaded rotors which rotate inside a close-fitting sleeve, drawing oil in at one end of the sleeve and driving it out at the other end.

The rotors of the IMO pump, which resemble worm gears, are shown in Figure 12-1. The inside diameters of the *spiral threaded portions* of the rotors are known as the *troughs* of the thread; the outside diameters or crests are known as the *lands*. The troughs and lands of adjacent rotors are so closely intermeshed that as they rotate, the meshing surfaces push the oil ahead of them through the sleeve, forming, in effect, a continuous seal so that only a negligible fraction of the oil that is trapped between the lands can leak back in the direction opposite to the flow.

The center rotor is power driven; its shaft is directly coupled to a 15-hp electric motor which drives it at 1750 rpm. The other two rotors, known as idlers, are driven by the center rotor which, through the intermeshing of its threads with the idlers, communicates the shaft power to the idlers and forces them to rotate in a direction opposite to the center rotor. The

h. The back-pressure tank, or *volume tank*, located in the control room and containing compressed air at a pressure of 10 to 25 psi, provides the air pressure on top of the oil in the main supply tank which keeps the entire system full of oil.

rotation of the center rotor is clockwise as viewed from the motor end of the coupling shaft, while the two idler rotors rotate counterclockwise.

135

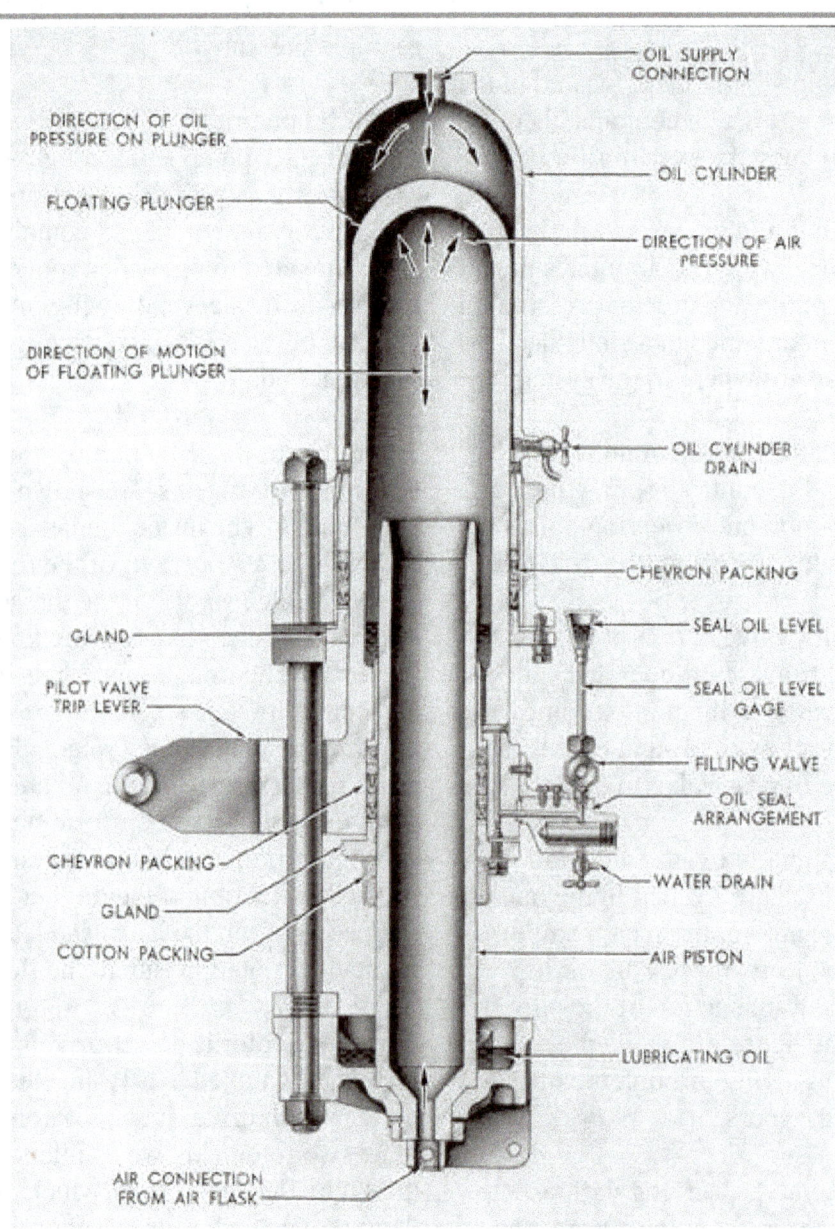

Figure 12-2 Hydraulic accumulator.

136

The end of the power rotor nearest the motor rotates in the guide bushing; the rotor shaft extends out through the end plate, where it couples to the shaft of the electric motor which drives it. Leakage around the shaft is prevented by five rings of 3/8-inch square flexible metallic packing which is held in place by a packing gland. Oil which leaks through the packing gland falls into the drip cup.

12B3. The main supply tank. Fluid is supplied to the pumps from the main supply tank. (See Figure A-9). The shape of this tank varies in different installations. Its total capacity is 50 gallons, but the normal supply maintained is only 30 gallons; the 20-gallon difference is an allowance made for discharge from the accumulator and thermal expansion of the oil.

When the system is operating, the fluid circulates through the power system, returning to the supply tank. However, the fluid will not remain in the supply tank for any length of time, but will be strained and again

pumped under pressure to the accumulator and the manifolds.

Glass tube sight gages mounted on the side of the reservoir, or supply tank, give minimum and maximum readings of the amount of oil in the tank. A drain line and valve near the bottom of the tank provide a means for draining water that may have accumulated there.

The back-pressure tank is connected by a length of pipe to the top of the supply tank (air inlet). It maintains an air pressure of 10 to 25 psi on the oil in the supply tank. This forms an air cushion between the top of the tank and the body of the fluid and maintains the system in a filled condition. An air relief valve set to lift at 40 pounds prevents the building up of excessive air pressures in the supply tank.

12B4. Accumulator. The 1,500-cubic inch air-loaded hydraulic accumulator is located in the pump room. (See Figure A-19.) Figure 12-2 shows a schematic view of the accumulator.

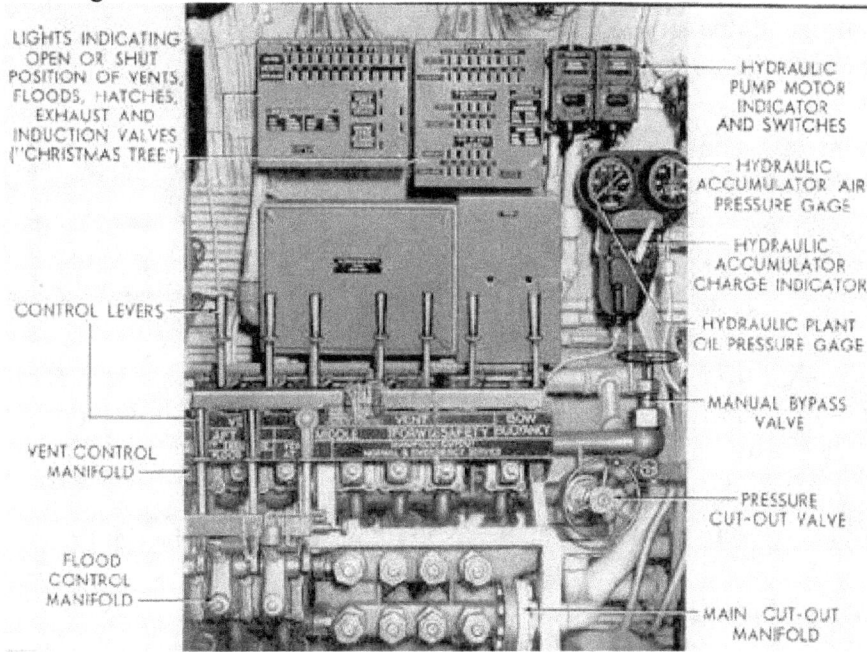

Figure 12-3. Main hydraulic control station.

137

The accumulator is essentially a hollow plunger free to move vertically within a stationary oil cylinder and over a stationary hollow air piston. The oil cylinder is connected to the pressure side of the manifold and the hollow air piston is connected to an air flask. The air flask is located in the pump room on the port side. The flask is charged through the accumulator air-loading manifold (located in the control room) from the high-pressure air system to a maximum of 1,950 psi to give a maximum oil pressure of 750 psi. The top of the movable plunger is therefore subjecting oil in the cylinder to a pressure caused by the air pressure with in the plunger. An indicator showing the position of the accumulator plunger is installed in the control room adjacent to the main manifold.

The accumulator performs the following functions:

a. It controls oil delivery to the hydraulic system from the hydraulic pump.
b. It maintains a constant pressure on the hydraulic system.
c. It provides a reserve supply of oil under pressure to permit the operation of gear when the pump is shut down and to supplement the supply of oil from the pump when several hydraulic gears are operated simultaneously.
d. It reduces shock to the system when control valves are operated.

12B5. Main hydraulic control station.

7. The manual bypass valve.
8. The pressure cutout valve,
9. The hydraulic accumulator charge indicator.
10. The Christmas Tree.

Thus, all units necessary to control the main hydraulic system are grouped in one place for efficiency and facility of operation.

12B6. Main cutout manifold. The main cutout manifold consists of eight valves, four of which are return valves on the upper row of the manifold, and four of which are supply valves on the lower part of the manifold.

The Supply valves from forward to aft, control the following:

a. Hydraulic service forward.
b. Emergency steering.
c. Emergency bow and stern plane tilting and normal bow plane rigging.
d. Hydraulic service aft.

The return valves from forward to aft, control the following:

a. Hydraulic service forward.
b. Emergency steering.
c. Emergency bow and stern plane tilting and normal bow plane rigging.
d. Hydraulic service aft.

12B7. Pilot valve. The pilot valve is used in the main hydraulic system to operate the automatic bypass valve by directing

Normal operation of the various hydraulically operated units of the vessel is controlled from the main hydraulic control station is the forward port corner of the control room. (See Figure 12-3.)

Here, in one group, are located:

1. The main cutout manifold.
2. The vent control manifold.
3. The flood control and engine air induction manifold.
4. The IMO pump stop and start push buttons.
5. The main plant oil pressure gage.
6. The hydraulic accumulator air pressure gage.

oil under pressure to the automatic bypass valve piston when the accumulator is fully charged, thereby opening the bypass, and then venting off this oil when the accumulator is discharged, allowing the bypass to shut again. It is mounted on or near the accumulator in such a way that the operating arm is actuated by a cam roller which is mounted on the accumulator plunger. Hydraulic fluid from the accumulator under pressure enters the valve at the supply port. As the accumulator is charged, the plunger moves downward, carrying with it the cam roller. As the plunger approaches the bottom of its stroke, the cam bears against the lower end of the pilot valve operating arm, pulling the piston down within the cylinder. In this position, the

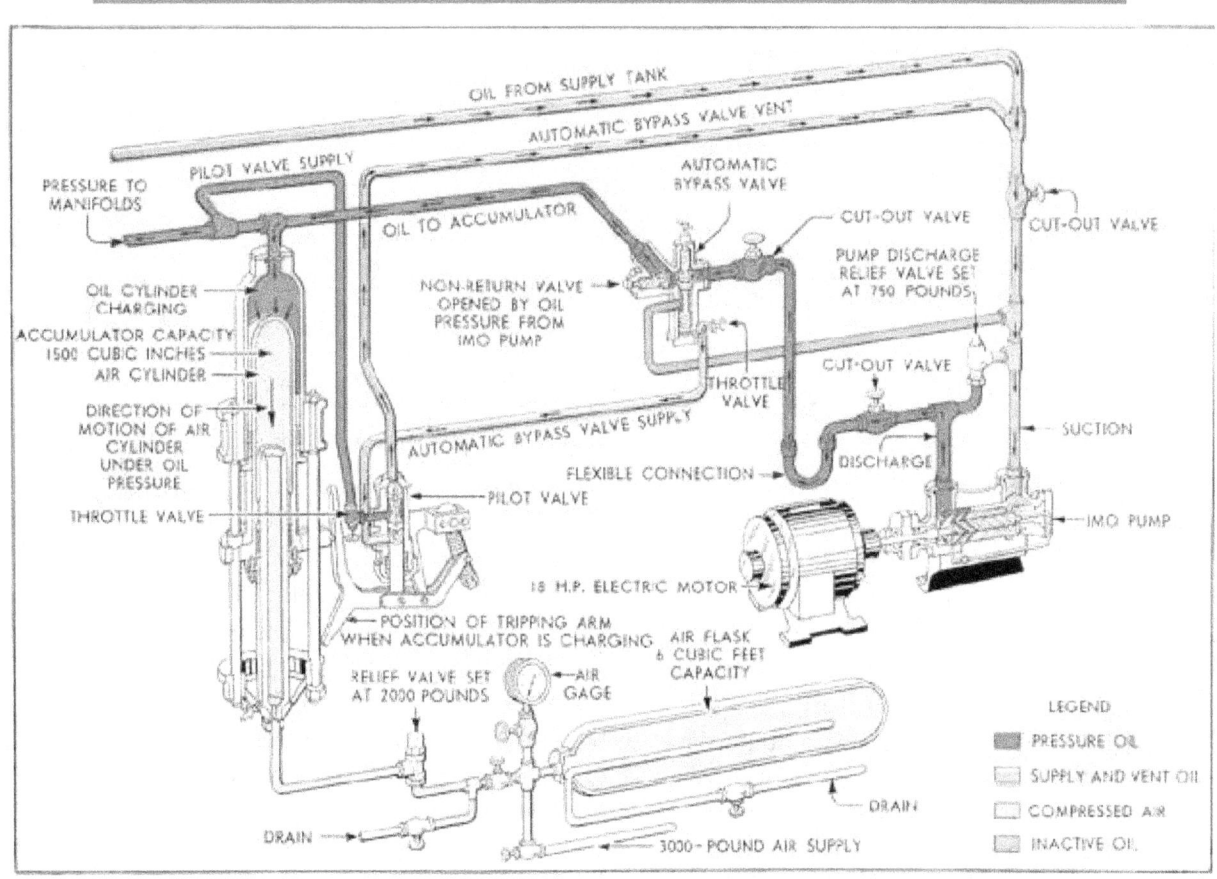

Figure 12-4, Charging the hydraulic accumulator.

139

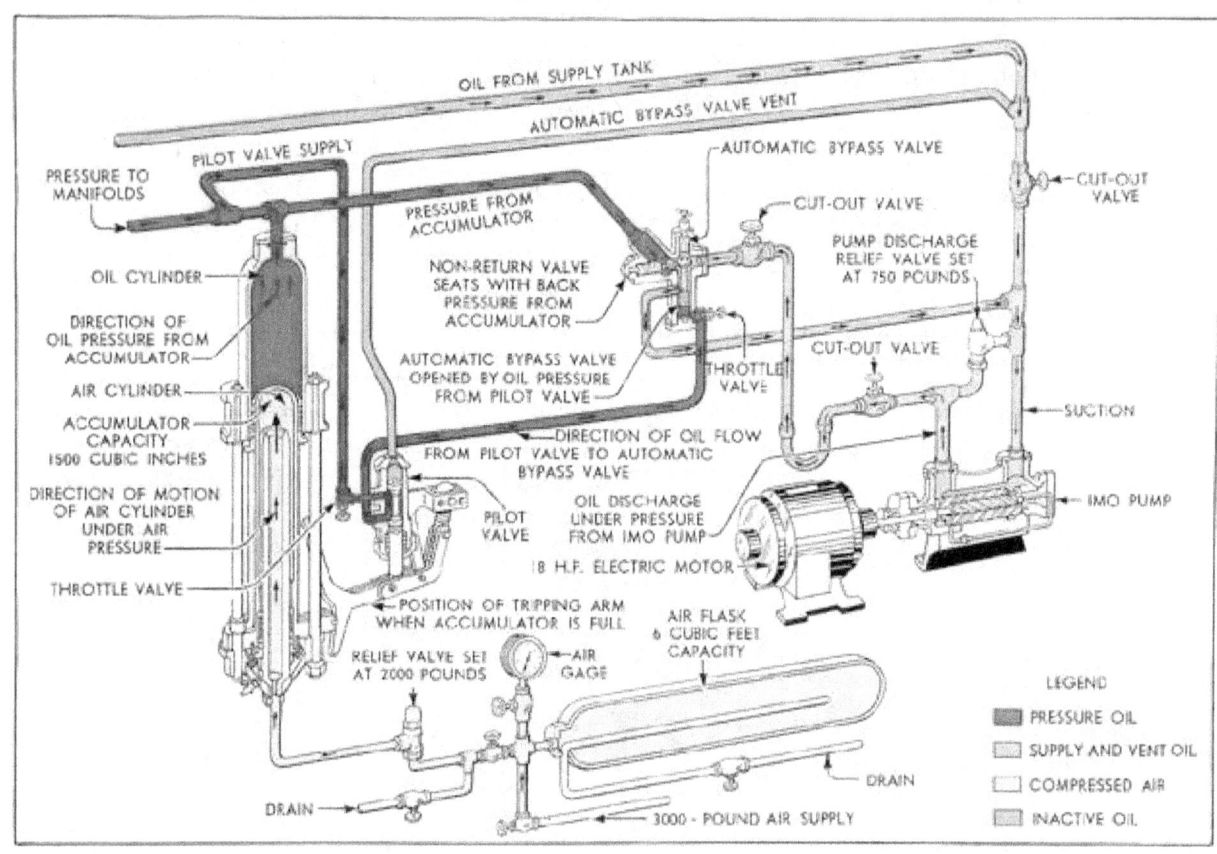

Figure 12-5. Accumulator discharging.

flat-milled surface cut along the side of the piston allows a column of oil to pass from the supply port through the port leading to the automatic bypass valve. (See Figure 12-4.)

This opens the automatic bypass valve, bypassing the pressure oil from the discharge side of the IMO pump back to the supply tank and allowing the nonreturn valve to seat. No more oil is delivered to the accumulator while the bypass piston, thus forcing the automatic bypass valve off its seat, and allowing the oil from the pumps to return to the supply tank. When this happens, there is not enough pressure to keep the nonreturn valve off its seat, so the disk valve spring returns the disk to its seated position, thus blocking the backflow of oil from the accumulator. Oil pressure from the accumulator also assists in the seating of the valve.

pilot valve remains in this position.

12B8. Automatic bypass and nonreturn valves. The automatic bypass and nonreturn valves are installed between the IMO pumps and the accumulator. There is one on each pump pressure line. The automatic bypass valve bypasses hydraulic oil when the accumulator is fully charged. The nonreturn valve prevents backflow of the oil from the accumulator to the pump.

As seen in Figure 12-4, the valve body contains two valve parts. One is the bypass valve which is held on its seat by the valve spring. The nonreturn valve is of the disk type which is also seated by a spring.

During those intervals when the accumulator is being charged, hydraulic oil is delivered by the pump into the automatic bypass and the nonreturn valve housing. The oil pressure unseats the spring-held nonreturn valve disk, and oil, under pressure, goes into the line to the accumulator. When the accumulator is fully loaded, the pilot valve is tripped and oil is directed to the automatic

When the oil charge in the accumulator is depleted by the use of oil to operate various units in the system, or by leakage, the plunger rises, causing the cam roller to bear against the upper end of the pilot valve operating arm, thus moving the pilot valve piston up until the land between the two flat-milled surfaces on the piston blocks off the supply port from the port leading to the automatic bypass valve. At the same time, the upper flat surface lines up the port with the escape port, venting the oil trapped under pressure in the line leading to the automatic bypass piston out through the port to a vent line, which bleeds into the main supply tank. This removes the pressure from underneath the valve piston of the automatic bypass, allowing the loading spring to reseat the automatic bypass valve and thus shut off the bypass line.

Immediately, oil under pressure from the IMO pump, once more directed against the underside of the nonreturn valve, opens this valve, allowing the oil to flow to the accumulator.

C. OPERATIONS

12C1. Starting the main plant. Following are the operations for starting operation of the main plant:

a. Check the supply tank for proper oil level.

b. Check the back-pressure for proper pressure in the supply tank.

c. See that the hand levers on the control manifolds are in the

e. Check the PRESSURE CUTOUT valve to see that it is open.

f. The manual bypass valve should be opened before starting the motor; after the motor has come up to speed, shut the manual bypass valve. This procedure is precautionary as the motor is not capable of properly starting and coming up to speed under full load.

NEUTRAL position.

d. Check the accumulator air flask pressure to see that the AIR TO ACCUMULATOR valve is open on the air-loading manifold.

12C2. Securing the main plant. Securing the main plant is accomplished as follows:

a. See that the hand levers are in NEUTRAL position on the control manifolds.

b. Stop the motor.

c. Open the manual bypass valve allowing oil to be drained from the accumulator to the supply tank.

12C3. Venting the system. When venting the system, vent all lines, valves, manifolds (except the air-loading manifold), accumulator, gages, control gears, and operating gears. Operating lines are vented by opening the vent valves at the operating gears. Vent valves in the operating lines that require venting are located abaft the diving station on the port side of the control room.

The system should be vented if it has not been in use for several days. The vents should be opened only when there is pressure on the lines.

12C4. Flood and vent control manifold. The main vent control manifold on the submarines built by the Electric Boat Company houses seven control valves instead of six as on the Portsmouth installations.

Reading from right to left, these seven valves operate the following vent valves:

1. Bow buoyancy tank.

The lines to the hydraulic operating cylinder are shut off so that if there is a break in the local circuit, oil will not leak out of it from the main system, and only the local circuit's oil will be lost.

The frame mounted on the manifold has notches cut into it for each valve position, into which the hand lever is firmly latched by a lateral spring. Once placed in any position, it cannot move unless purposely moved by the operator.

Each of these control valves operates a flood or vent valve, at some point remote from the manifolds, by directing a column of pressure oil to one side or the other of a hydraulic unit cylinder whose piston is connected, through suitable linkage, to the valve operating mechanism. All MBT vent valves and the safety tank and bow buoyancy vent valves are hydraulically operated.

The operating gear consists essentially of a hydraulic unit cylinder and suitable linkage connecting it to a vertical operating shaft which opens and shuts the vent. Fluid under pressure is admitted from the control valve into the hydraulic operating cylinder. As the piston head moves, it actuates the crank shaft. This moves the cam, which, bearing against the groove in the slotted link, causes it to push up or pull down

2. Safety tank.

3. MBT Nos. 1 and 2.

4. FBT Nos. 3 and 5.

5. FBT No. 4.

6. MBT No. 6.

7. MBT No. 7.

Reading from right to left, the flood and induction valve levers are:

1. Engine induction and ship's supply outboard valve.

2. Negative flood.

3. Safety flood.

Each valve has four positions which are shown on the indicator plates next to the hand levers:

1. SHUT, which closes the vent.

2. OPEN, which opens the vent.

3. HAND, which bypasses the oil allowing hand operation.

4. EMERGENCY, which shuts off the lines to the hydraulic operating cylinder.

on the flat link, thereby moving the crosshead up or down. Into the top of the crosshead is screwed the lower end of the operating shaft. This shaft goes up through a packing gland in the pressure hull to the superstructure, where the mechanism that opens and shuts the vent is located.

12C5. The hydraulic flood valve operating gear. The flood valves on the safety and negative tanks are hydraulically operated. The crossarm and hand grips are for hand operation in case of failure of the hydraulic power.

It is essential to understand that the main piston rod and the tie rods are all rigidly yoked together through the crosshead. Impelled by the hydraulic pressure against the piston head, all three rods move inward or outward as one solid piece. To open the valve, hydraulic fluid from the control valve is admitted into one end of the cylinder

moving the piston head outward. The motion is communicated through the crosshead. The tie rods, screwed rigidly into this crosshead, are pushed outward; the outboard connecting rods, through the crank, push the operating shaft out, opening the flood valve. Return oil, meanwhile, flows out of the opposite end of the cylinder back to the control valve.

To shut the valve, the flow of hydraulic fluid is reversed, pushing the

hydraulic fluid only to the lower ends of the cylinders. No oil is present on top of the piston heads except that which leaks past the piston from the pressure side. Overflow lines and a settling tank located in the conning tower are provided to catch any oil that may leak up past the piston heads.

To lower the periscope, the lines from the ports at the lower ends of the cylinders are simply opened to the return line, and the periscope and

button inward.

12C6. The periscope. A pair of hydraulic cylinders is bracketed into the periscope fairwater, at the top of the conning tower. The piston heads and piston rods are bolted to a yoke which carries the periscope; in other words, the pistons and periscope are rigidly connected together and travel as a unit. As the pistons are raised by admitting hydraulic pressure to the undersides of the piston heads, the periscope extending through the center of the fairwater slides up from its well and is projected upward.

A distinctive feature of this type of hoist is the fact that the control valve admits

pistons are allowed to descend by their own weight, forcing the oil out of the cylinders into the return line.

12C7. The vertical antenna hoist. The vertical antenna hoist need not be discussed in detail, as it is almost identical to the periscope hoist in arrangement, structure, and operating principles.

In addition to the automatic trip arrangement for avoiding the hard stop at the top of its travel, the vertical antenna hoist also has a dash-pot arrangement and a piston head with tapered grooved cut toward its underside, which help to bring it to an easy stop at the bottom.

D. FORWARD AND AFTER SERVICE LINES

12D1. General arrangement. There are two sets of hydraulic lines extending from the main cutout manifold to both ends of the submarine. These lines, known as the foreward and after service lines, furnish power to a miscellaneous group of hydraulically operated submarine equipment; specifically, these lines service the following apparatus:

a. The after service lines supply power for the operation of:

1. Main engine outboard exhaust valves (hydropneumatic on latest installations).
2. Outer doors of the four after torpedo tubes.
3. Periscopes and vertical antenna hoists (latest installations).

3. Two echo ranging and sound detection devices (raise or lower).
4. Outer doors of the six forward torpedo tubes.

Hydraulic pressure is distributed to the service lines at the main cutout manifold by two valves. One line is marked *Service forward*, the other line is marked *Service aft*. The return lines terminate in two similarly named valves on the main cutout manifold.

12D2. Torpedo tube outer door mechanism. The torpedo tube outer doors are hydraulically operated as separate units from the fore and aft service lines. There are ten torpedo tubes in all, six forward and four aft.

The outer door-operating mechanism consists essentially of the hydraulic

b. The forward service lines supply power for the operation of:

1. Bow plane rigging.
2. Windlass and forward capstan.

cylinder, piston and power shaft, the control valve and operating handle, and a jack screw for hand operation. All parts are mounted on the torpedo tube itself and controlled from its breech.

The hydraulic cylinder contains a piston that is moved by hydraulic power. It is connected rigidly to the power-operating shaft, the motion of which opens or shuts the outer door. The hydraulic power is directed to one side or the other of the hydraulic cylinder by the control valve. This allows flow of hydraulic power from the supply side of the forward or after service lines and feeds it back to the return side. The control valve is operated by the operating handle, a push-pull arrangement which slides in and out lengthwise through the ready-to-fire interlock tube. The operating handle is connected to the control valve by the inner slide which is attached to the control valve linkage by the operating lug.

Safe operation of a torpedo tube is a *delicate and complicated process*, involving a number of different conditions which cannot be allowed to occur simultaneously. For example, it is obvious that when the outer door is opened to the sea, the inner door must be locked shut and vice versa; the tube must not be made ready-to-fire unless different interlocks are properly engaged. For hand operation of the outer doors, a hand-operating shaft is provided, with a squared end, over which an operating crank fits. This turns the hand shaft driving gear. This

position so that the fluid trapped in the hydraulic cylinder will not act as a hydraulic lock against the motion of the piston.

The operating handle therefore has three positions: *OPEN* (handle pulled all the way out toward the operator), in which the power operating shaft, moved by hydraulic power, will open the outer door; *SHUT* (handle pushed in all the way away from the operator), in which the power-operating shaft will shut the outer door; and *HAND* (handle in intermediate position), in which the lines from the hydraulic cylinder are by-passed through the control valve.

12D3. Echo ranging and detecting apparatus. The echo ranging and detecting apparatus is contained in a metal sphere (called the sound head) fixed to a cylindrical tube which is extended downward through an opening in the underside of the vessel in much the same way that the periscope is extended upward through the top. The tube is hydraulically operated by power from the forward service line of the main hydraulic system.

The hydraulic part of the apparatus consists essentially of three hollow tubes, one within the other, so arranged that the two inside tubes act as a stationary piston fixed to the frame of

gear is meshed with the jack nut, which in turn is threaded into the threaded portion of the power-operating shaft. Therefore, as the jack nut is turned, the power-operating shaft travels through it, opening or shutting the outer door. In order to operate this by hand, the control valve must be in the hand

the vessel, while the outermost tube, actuated by hydraulic pressure, acts as a movable cylinder which slides up and dozen over it, raising or lowering the sound head. A control valve directs the oil pressure to one side or the other of the piston head to raise or lower the cylinder.

A hand pump is installed in the lines for hand operation.

E. EMERGENCY STEERING AND PLANE TILTING SYSTEMS

12E1. General. The steering and plane tilting operations are usually performed by their own individual hydraulic systems. To insure against failure, it is possible to use the pressure in the main hydraulic system to power

the gear that actuates the rudder and the planes. In the main hydraulic system, this is accomplished by connecting supply and return lines from the other systems to the main cutout manifold.

13
THE STEERING SYSTEM

A. INTRODUCTION

13A1. General. The rudder of the submarine is moved by hydraulic power. Under normal operation, the steering system has its own source of power, a motor-driven size 5 Waterbury A-end pump, and is, therefore, except in emergencies, completely independent of the main hydraulic system described in Chapter 12.

The principal control units are assembled in the steering stand, located in the control room. However, since there is a steering wheel in the conning

2. *Hand*, in which the hydraulic power is developed in the steering stand pump by the direct manual efforts of the steersman.
3. *Emergency*, in which the hydraulic power is supplied by the main hydraulic system.

It should be emphasized that the rudder itself is moved by hydraulic power in all three cases; the only difference between these methods is in the manner in which *the power is developed.*

tower connected to the steering stand controls by a shaft, the submarine can be steered either from the control room or from the conning tower. To allow for every contingency, the steering system is so planned that three different methods of steering are available, based on three different sources of hydraulic power. They are designated as follows:

1. *Power*, in which the hydraulic power is independently developed by a motor-driven pump belonging to the system itself.

Emergency power is used only when the normal power (called simply *Power*) fails. *Hand* power is used only when silent operation of the submarine is necessary to avoid detection by enemy craft, or when both the normal *Power* and the *Emergency* power from the main hydraulic system have failed.

The submarine can be steered by all three methods from either the control room or the conning tower.

B. DESCRIPTION

13B1. General arrangement. The steering system as a whole is shown in Figure A-20. The system may conveniently be thought of as divided into four principal parts:

a. The normal power supply system, comprising a Waterbury size 5 A-end pump, the motor which drives it, the control cylinder, and the main manifold.

b. The steering stand, comprising the main steering wheel, emergency handwheel, steering stand pump, pump control lever, change valve, emergency control valve, conning tower connecting shaft, and a clutch.

c. The main cylinder assemblies, comprising the cylinders and plungers and the mechanical rudder-angle indicator.

d. The rudder assembly, comprising the connecting rods and guides, the crosshead, on the rudder itself.

13B2. Detailed description. *The normal power supply system. The Waterbury speed gear.* The actuation of the various hydraulically operated units on board a submarine often requires great precision of control and the transmission of power at variable speeds and pressures, without any sharp steps or graduations. The hydraulic machine used for many of these operations is the Waterbury speed gear, a mechanism which furnishes instant, positive, and accurate hydraulic power transmission.

The Waterbury speed gear may be used as a pump (converting rotary mechanical motion into hydraulic fluid displacement) or, with one important modification of internal structure, as a hydraulic motor (converting hydraulic fluid displacement into rotary mechanical motion).

The type of Waterbury speed gear generally used as a pump is designated as a Waterbury A-end speed gear (Figure 13-1). The type used exclusively as a hydraulic motor is designated as a Waterbury B-end speed gear or Waterbury B-end hydraulic motor. The A-end type is in one special installation used as a hydraulic motor, but, since this is not generally the case, it will be convenient to describe the A-end type primarily as a hydraulic pump.

A-end and B-end speed gears are often used together to form a pair of power transmission units separated by any required length of hydraulic piping to suit the particular installation needs. So used, they receive rotary mechanical motion from an electric motor at one point and transmit it as fluid displacement any required distance, where it is reconverted into rotary motion with a positiveness and fineness of control that could not be achieved by the use of electric motors alone.

Waterbury A-end speed gear, used in the submarine hydraulic system primarily as a pump, is designated as size 5-A. Two sizes of B-end motors are used, designated respectively as 5-B and 10-B.

The Waterbury A-end pump is operated by a rotating shaft which may be driven either by an electric motor or by hand. Three motor-driven and three hand-driven Waterbury A-end pumps are used in a submarine: one of each type in the, steering system, stern plane system, and bow plane system, respectively. In operation by normal power, the two types are used in each system as a *team*; the motor-driven unit

1. A socket ring, which holds the ball sockets of the seven or nine piston connecting rods arranged in a circle around the driving shaft.

2. A cylinder barrel, in which are bored the seven or nine corresponding cylinders.

3. A tilting box, which alters the angle and direction of the socket ring with respect to the cylinder barrel.

The socket ring and cylinder barrel are mounted on the drive shaft so that they rotate together. The socket ring is sea arranged that it can be made to rotate either parallel to the cylinder barrel or at an angle to it. Connected to the tilting box is a control shaft extending through the pump casing, which, when pushed up or down, determines the angle and direction of the tilting box.

Reference to Figure 13-1 will help to clarify the manner in which pumping action is obtained. The socket ring rotates within the tilting box on the radial and axial thrust bearing. As long as the tilting box is maintained in the vertical position, the socket ring and cylinder barrel rotate parallel to each other, and there is no reciprocating motion of the pistons within the cylinder barrel. However, when the tilting box is tilted in either direction away from the vertical, the socket ring no longer rotates in the same plane, as the cylinder barrel. This means that as a ball socket on the socket ring reaches that point in its rotation which is closest to the barrel, the piston belonging to it will be driven down into the corresponding cylinder, and then, as this same ball socket recedes to the point farthest away from the barrel, the piston transmits oil or the power actuation of the system, while the hand-driven unit, fitted with a large handwheel and designated as a telemotor, or steering stand pump, transmits oil to a control cylinder to provide fine control of the output of the motor-driven units. The hand-driven unit is also used, alternately, to operate the system by hand whenever it is desired not to use the motor-driven pump.

Although the Waterbury A-end speed gear is activated by rotary motion, in principle it is actually a reciprocating multiple piston type of pump. It consists of a casing containing three basic elements:

will again be withdrawn.

The diagrams on the lower part of Figure, 13-1 showing the tilting box tilted away from the vertical, and illustrate the course of a single piston, whose motion we are able to follow as the socket ring turns through half a cycle (180 degrees).

As the piston rises to its uppermost position, it occupies a progressively smaller space in the cylinder until it reaches the point at which the socket ring and barrel are farthest apart. The partial vacuum which is produced in the chamber by the outward movement of the piston draws the fluid into the cylinder by suction.

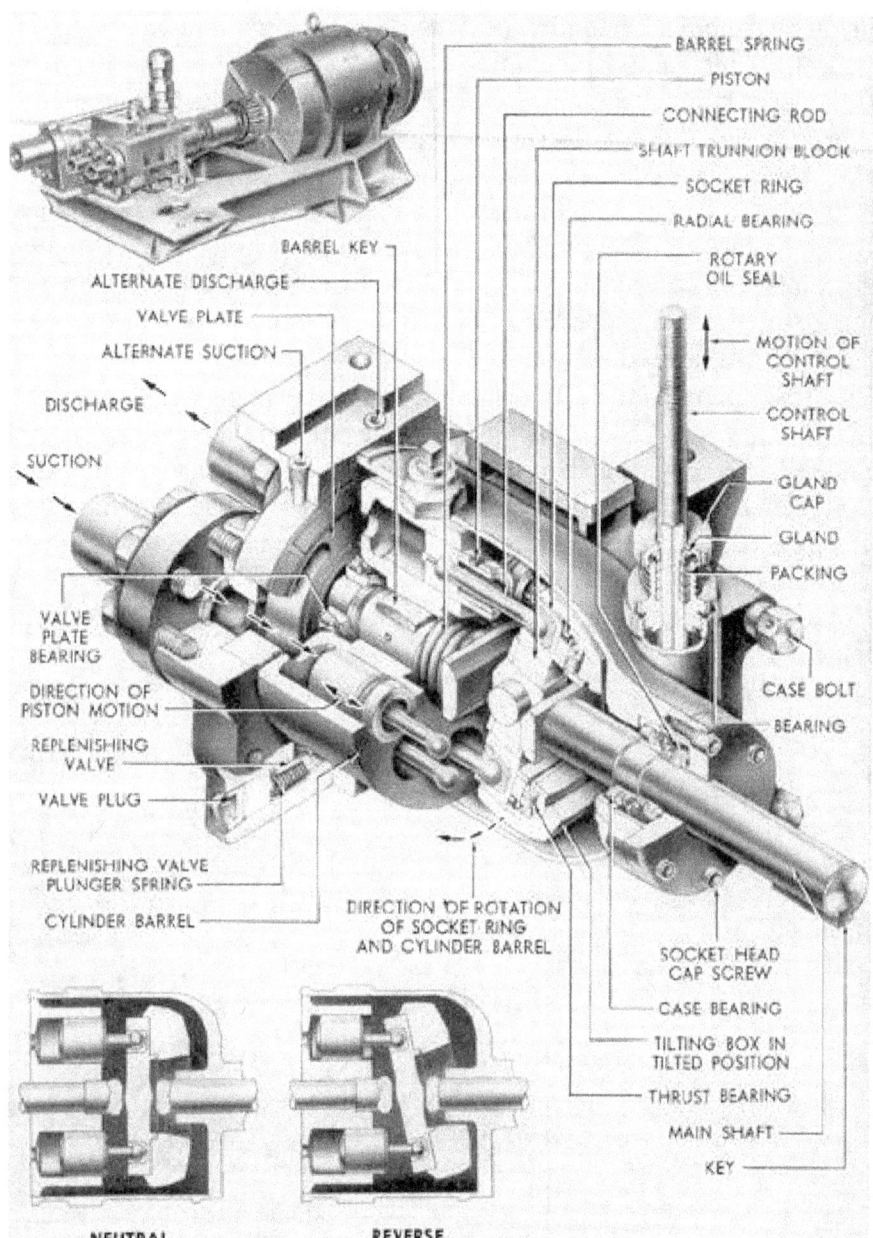

BARREL SPRING
PISTON
CONNECTING ROD
SHAFT TRUNNION BLOCK
SOCKET RING
RADIAL BEARING
ROTARY OIL SEAL
MOTION OF CONTROL SHAFT
CONTROL SHAFT
GLAND CAP
GLAND
PACKING
CASE BOLT
BEARING
SOCKET HEAD CAP SCREW
CASE BEARING
TILTING BOX IN TILTED POSITION
THRUST BEARING
MAIN SHAFT
KEY

BARREL KEY
ALTERNATE DISCHARGE
VALVE PLATE
ALTERNATE SUCTION
DISCHARGE
SUCTION
VALVE PLATE BEARING
DIRECTION OF PISTON MOTION
REPLENISHING VALVE
VALVE PLUG
REPLENISHING VALVE PLUNGER SPRING
CYLINDER BARREL

DIRECTION OF ROTATION OF SOCKET RING AND CYLINDER BARREL

NEUTRAL   REVERSE

Figure 13-1. Waterbury speed gear.

147

In the intermediate position, the piston returns into the cylinder and begins to displace the fluid accumulated there. At its lowest point, the piston occupies almost the entire cylinder. The expulsion of the fluid through the discharge port is now complete. The piston again rises from this position for the suction stroke. The repetition of

The pump control shaft enters at the bottom, connected to the tilting box. The centering spring and its actuating spindle, against which the top end of the pump control shaft bears, are contained in the tall, pipe-like housing screwed onto the top of the power-driven Waterbury A-end pump.

these movements in sequence by all of the pistons results in a smooth nonpulsating flow of hydraulic fluid.

In normal operation, the hydraulic power used by the steering system is developed by a Waterbury size 5 A-end pump. It is driven by a 1.5-hp electric motor at a constant speed of about 440 rpm. The pump turns in a clockwise direction as viewed from the, motor end of the shaft. The pump's speed is constant; only the direction and angle of the tilting box change. It is these that determine the amount of oil that is pumped into the system to move the rudder and the direction in which it is pumped.

b. *The control cylinder.* The function of the control cylinder is to translate the movement of the main steering wheel, as the steersman turns it left or right, into a corresponding upward or downward motion of the control shaft, thereby changing the position of the tilting box in the motor-driven Waterbury pump. This, in turn, varies the stroke of the pistons inside the motor-driven pump. It also determines the quantity and direction. of flow of the oil that is pumped to the main rams. In this manner it controls the output of the motor-driven Waterbury pump in obedience to the actions of the steersman when steering by normal *power.*

The control cylinder assembly consists of a pair of small hydraulic cylinders opposed and axially in line, having in common a single plungers which slides between and through the cylinders. Bell-crank linkage connects this plunger to the tilting box.

13B3. The steering ram cutout manifold. The steering ram cutout manifold consists of a multiple-port housing containing nine valves built into the body, and eight ports which connect the main rams to the sources of hydraulic power.

The manifold is so arranged that the four center valves are power cutouts to the port and starboard rams from the main steering pump. The forward set of two valves and the after set of two valves are hand and emergency cutouts to the port and starboard rams when the power is furnished from the control room. A bypass valve at the top central part of the manifold, if opened, would bypass the main steering pump by connecting both sides of the pump together. This bypass normally is shut.

The manifold has two connections at the top which connect the manifold with the motor-driven Waterbury A-end pump. Of the lower four connections of the manifold, the two in the center are connections to the starboard ram. The remaining two connections, one forward and one aft on the lower part of the manifold, are hand and emergency connections from the control room.

The connections from the manifold to the port ram are at the foremost and aftermost part of the manifold. All the valves have attached name plates indicating their purpose.

The main cylinder ram assemblies, usually referred to as *the rams* (port and starboard), transform hydraulic power into mechanical power to move the rudder. Each consists essentially of a pair of hydraulic cylinders opposed and

On all later classes of submarines, the control shaft that extends through the Waterbury A-end power-driven pump has the centering spring attached to one end of the control shaft and the control cylinder on the opposite end.

axially in line, having in common a plunger or ram that slides between and through them and a hydraulic port at each end, into which oil is admitted to move the rams forward or aft. The plunger has at its center a heavy yoke forged

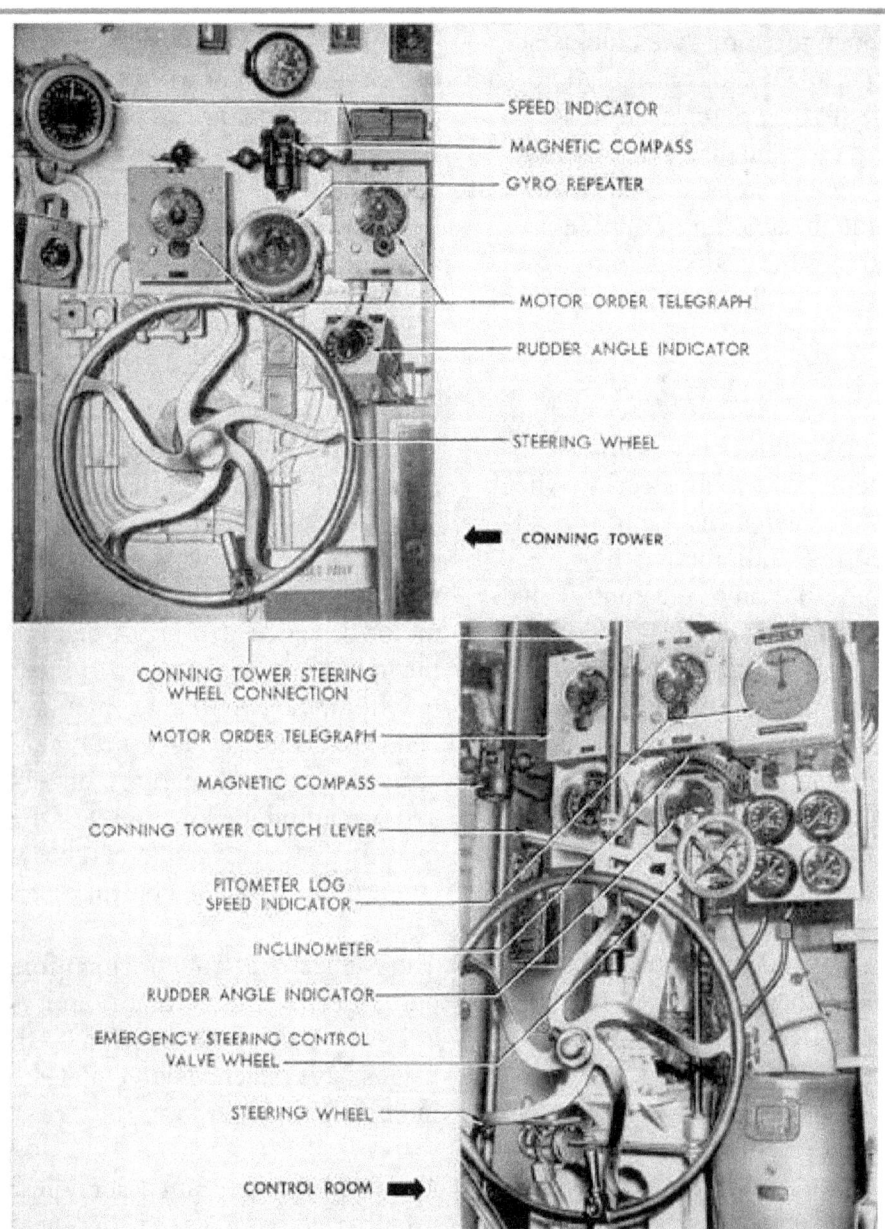

SPEED INDICATOR
MAGNETIC COMPASS
GYRO REPEATER
MOTOR ORDER TELEGRAPH
RUDDER ANGLE INDICATOR
STEERING WHEEL
CONNING TOWER

CONNING TOWER STEERING WHEEL CONNECTION
MOTOR ORDER TELEGRAPH
MAGNETIC COMPASS
CONNING TOWER CLUTCH LEVER
PITOMETER LOG SPEED INDICATOR
INCLINOMETER
RUDDER ANGLE INDICATOR
EMERGENCY STEERING CONTROL VALVE WHEEL
STEERING WHEEL
CONTROL ROOM

Figure 13-2. Steering stand.

integrally with it; the yoke has a hole

pressures developed by the motor-driven

drilled in it to take the inboard connecting rod which is locked into it at this point by heavy lock nuts, one on each side of the yoke. The inboard connecting rod slides through the bearings. Oil leakage past the plunger is prevented by the packing. The entire ram assembly is bolted to the framework through the brackets.

Mounted at the forward end of the ram is the mechanical rudder-angle indicator pointer showing the angle of rudder deflection on the indicator dial, which is graduated in degrees. An electrically operated rudder angle transmitter is located on the other ram. It transmits the angle of deflection electrically to a rudder angle indicator on the instrument hoard in the control room.

13B4. The steering stand. The hydraulic power that moves the rudder is directed by the steersman from the steering stand, an assembly which contains the control equipment for all three methods of steering, *Power, Hand, and Emergency.* (See Figure 13-2.)

a. *The steering stand pump.* Since, in operation by normal power, it is the direction of the motor-driven Waterbury A-end pump tilting box that determines which way the rudder moves, and since the position of this tilting box is controlled by the movement of oil in the control cylinder, it is clear that to steer the submarine, some device is needed which will drive that oil one way or the other as desired. The mechanism must be one that will respond readily to the steersman's touch, yet control accurately the powerful

Waterbury A-end pump. Such a device is the *steering stand pump*, the steering stand's main unit. The steering stand pump is actually a hand-operated Waterbury A-end pump. A bracket is fitted externally to it and the pump control shaft so that its tilting box always tilts in the same *direction*, though its *angle*, that is, the *degree* of tilt, may be changed. Consequently, the flow of oil depends solely on which way its shaft is rotated. If a large handwheel is fitted to this shaft, the ports of the pump connected to opposite ends of the control cylinder, turning the wheel left or right, will then pump oil to one or the other end of the control cylinder, which in turn tilts the tilting box in the motor-drivers Waterbury A-end pump, thus moving the rudder left or right. Therefore, turning a wheel fitted to the shaft of the steering stand pump will steer the submarine.

b. *The main steering wheel.* This wheel is mounted vertically at the after end of the steering stand. It is used for both POWER and HAND steering.

As hand steering requires greater effort, a retractable spring handgrip is built into the rim. During power steering, this handgrip may be kept folded in. A spring-loaded locking pin is built into the hub; when pulled out, it allows the main steering wheel to be disengaged from its shaft.

This is provided to prevent the main wheel from spinning heedlessly when the submarine is being steered from the conning tower.

C. OPERATIONS

13C1. Power steering. When steering by power, (See Figures 13-3 to 13-7.) the following conditions are obtained:

a. The change valve in the control room is set for *power* steering.

b. The steering stand pump stroke control lever may be in any of the possible positions. Experience has indicated, however, that the most satisfactory position is with the pump at approximately three-quarters of a stroke.

c. The main steering motor is running.

To illustrate the operation of the steering gear when steering by *power*, assume that it is desired to move the rudder from amidship to hard over left rudder. The steersman turns the steering wheel to the left, thereby turning the shaft of the steering stand pump which delivers oil through the change valve and one of the control cylinder lines to the after control cylinder; and oil from the forward control cylinder is forced back through

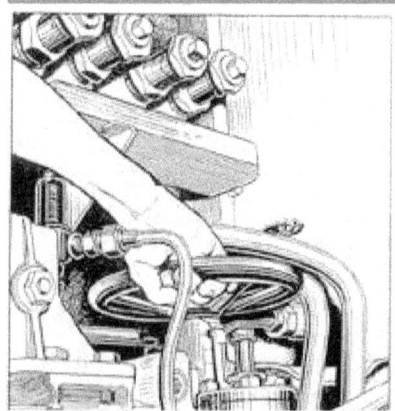

Figure 13-3. Change valve.

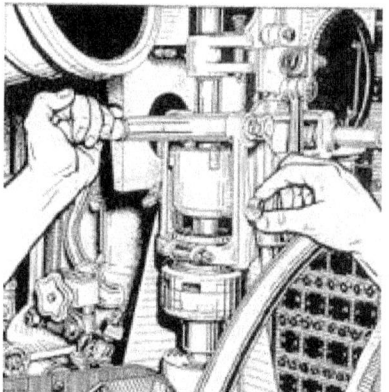

Figure 13-5. Shifting steering control.

the other control cylinder line to the suction side of the pump.

Delivery of oil to the after control cylinder moves the control ram forward, thereby moving the main pump tilting box control shaft downward from neutral toward full stroke. This puts the tilting box in a position to deliver oil from the port side of the pump through the relief and cutout manifolds. Oil from the manifolds enters the lines to the forward starboard ram and the after

For the purpose of maintaining the pump control shaft in neutral position when it is desired to hold the rudder angle constant, a spring-loaded centering device is mounted adjacent to the pump. This device consists of a compression spring enclosed in a cylinder and mounted on a spindle in such a way that if the spindle is moved in either direction, the spring is compressed and tends to return the spindle to its normal position. The spindle is connected to a lever mounted on the rocker shaft which operates the levers to the pump and control cylinders

port ram, moving the rudder to the left, while return oil from the forward port and after starboard ram is delivered to the starboard side of the pump or the suction side.

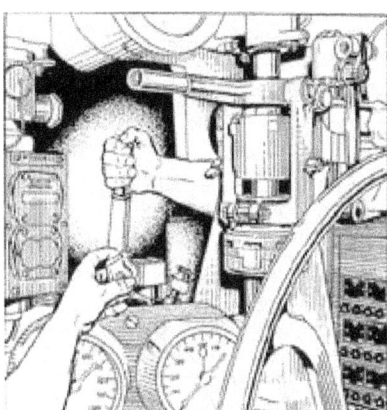

Figure 13-4. Stroke adjuster.

respectively.

When the desired position of the rudder is reached, the steering wheel must be brought

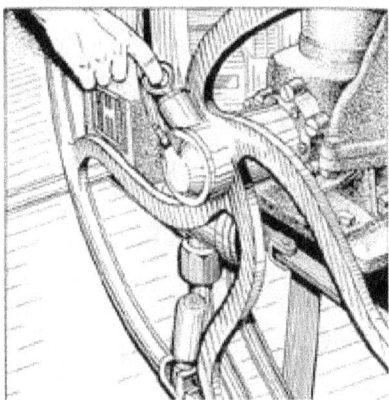

Figure 13-6. Steering wheel.

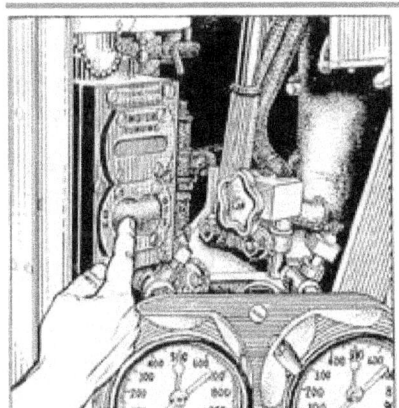

Figure 13-7. Starting control.

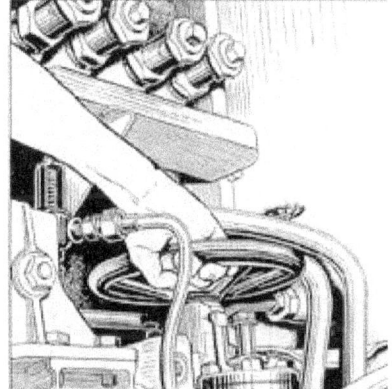

Figure 13-9. Change valve.

back to its original position to stop rudder movement, since there is no follow-up mechanism in this steering gear.

The power steering gear is protected by two relief valves, one installed in either side of the main relief manifold.

13C2. Hand steering. When steering by hand, (See Figures 13-8 to 13-12.) the following conditions are obtained:

is set in its aftermost position in order to obtain a maximum delivery of oil and therefore maximum speed of rudder travel under the condition of hand steering.

Again, assume that it is desired to move the rudder from amidships to the hard over left position. The steering wheel is turned left. Oil is delivered by the steering stand pump directly to the forward starboard ram and after port ram. The rudder moves to the left. Oil from

a. The main Steering pump aft is stopped.

b. The change valve in the control room is set for *hand* operation.

c. The steering stand pump stroke lever

Figure 13-8. Starting control.

the after starboard ram and the forward port ram returns to the suction side of the steering stand pump. The rudder moves so long as oil is delivered to the rams

Figure 13-10. Stroke adjuster.

152

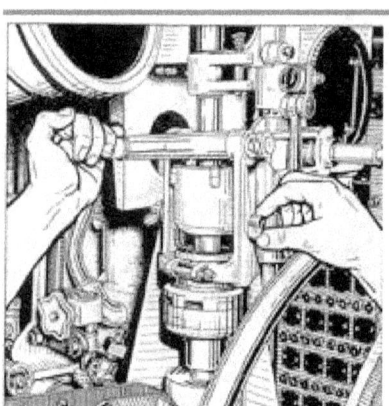

Figure 13-11. Shifting steering control.

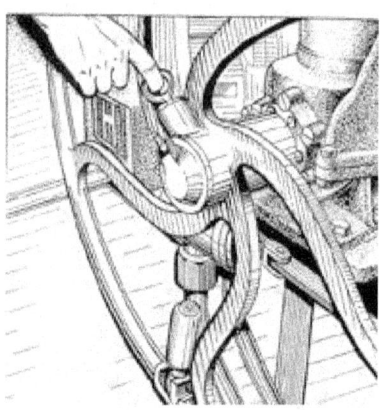

delivered to the forward port and after starboard rams while at the same time, oil is returned from the after port and forward starboard rams through the control valve to the return side of the main cutout manifold. Movement of the control valve handwheel for left rudder causes the oil to be delivered to the after port and forward starboard rams while, at the same time, oil is returned from the forward port and after starboard rams through the control valve to the return side of the main cutout manifold.

When steering by *emergency* power, the change valve should be set in the *emergency*

Figure 13-12. Steering wheel.

by turning the steering wheel and thus driving the steering stand pump.

13C3. Emergency steering. Provision is made for steering by direct delivery of oil to the main rams from the main hydraulic system. Oil is delivered from the main cutout manifold to the steering stand. The emergency steering control valve on the steering stand is a piston type control valve. Oil returns from this valve to the return and low-pressure side of the main cutout manifold. Movement of the control valve handwheel for right rudder causes the oil under pressure from the main cutout manifold to be

Figure 13-13. Starting control.

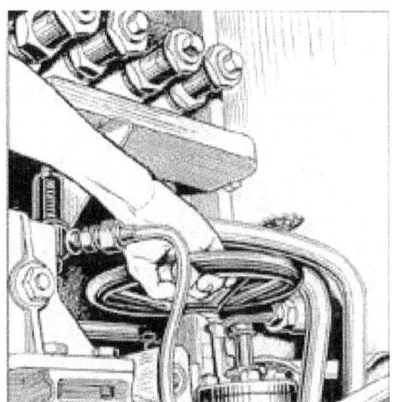

Figure 13-14. Change valve.

153

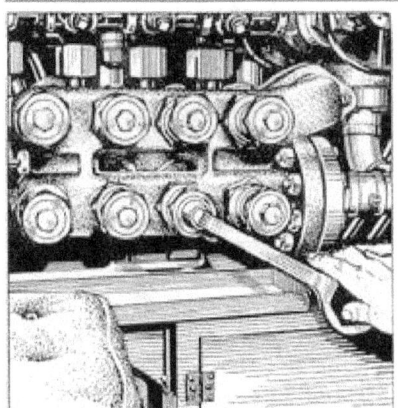

Figure 13-15. Main cutout manifold.

position. The emergency cutout valves in the hand and emergency cutout manifold should be opened and the hand cutout valves should be shut. When the desired position of the rudder is reached, the handwheel must be

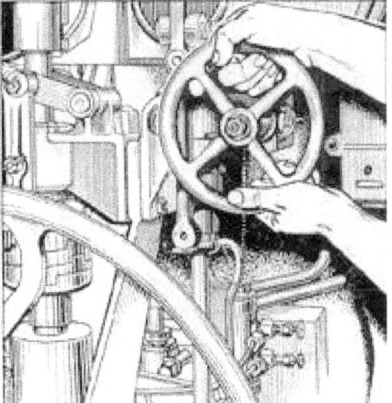

Figure 13-17. Emergency steering wheel.

conning tower, the handwheel control (in control room) must be disengaged. A clutch is provided for this purpose and must be engaged except when the link is connected for emergency steering from the conning tower. To steer by emergency from the control room, this

brought back to neutral to stop rudder movement and to hold the rudder in the desired position. Arrangement is provided to connect the emergency control valve lever to the vertical steering shafting by a removable link, thereby making it possible to steer by the emergency system from the conning tower.

When emergency steering from the

Figure 13-16. Emergency steering pin.

removable link is not connected and the emergency steering control valve is moved by the handwheel. A locking pin is provided to hold the control valve in the neutral position when emergency steering is not being used. (See Figures 13-13 to 13-20.)

The electrical rudder angle indicating system is of the selsyn type. The rudder

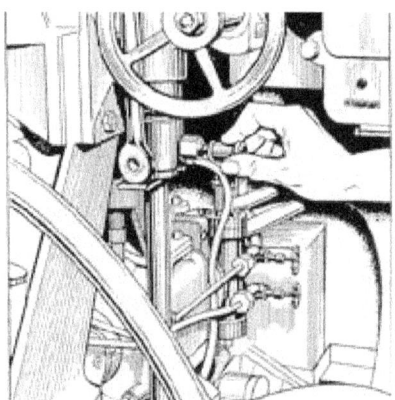

Figure 13-18. Emergency control valve lock.

154

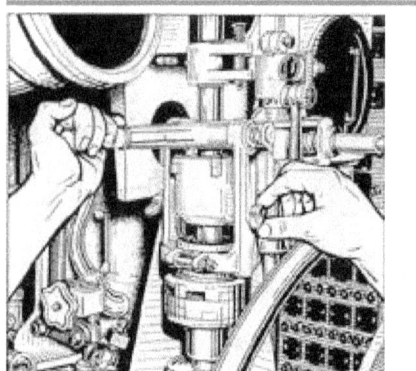

Figure 13-19. Shifting steering control.

angle transmitter is located in the after torpedo room on the port side and is driven through a rack and pinion from the port steering ram connecting rod. There is one rudder angle indicator in each of the following locations: the bridge, the conning tower

Figure 13-20. Emergency connecting link.

steering station, the control zoom steering station, the control room diving station.

A mechanical rudder angle indicator, driven also from the port steering ram connecting rod, is located in the after torpedo room.

155

14
BOW AND STERN PLANES SYSTEMS

A. INTRODUCTION

14A1. General. Hydraulic power is used to tilt the bow and stern planes. Each system (bow and stern planes) has its own power supply system. Except in emergencies, the power facilities in each system are adequate for its own individual operation without reliance on power from the main hydraulic system.

The control units for diving and rising are assembled in a diving control stand located in the control room. There is a set of controls for stern plane tilting, a set for bow plane tilting, and control valves for bow plane rigging. The control panel has diving indicators, gages, and motor switches.

Three methods of plane tilting are available at the control panel, based on three different sources of hydraulic power. They are designated as follows:

1. *Power,* in which the power is independently developed in each plane tilting system by the motor-driven Waterbury A-end pump belonging to that system.

2. *Hand,* in which the power is developed in the diving control stand pump, connected to each system, by the manual efforts of the diving stand operator.

3. *Emergency,* in which the power is obtained from the main hydraulic system.

Emergency is used only when normal power fails. Hand power is employed when the other two power sources are, inoperative, or when silent operation of the submarine is necessary to avoid detection by the enemy.

In addition to bow and stern plane tilting, this chapter also contains a description of bow plane rigging. A schematic view of the bow and stern planes systems and their associated equipment is illustrated in FigureA-21.

B. BOW AND STERN DIVING GEAR

14B1. Bow planes. Each plane is carried on a separate stock. The mechanism for tilting planes consists of a hydraulic cylinder and piston arrangement connected by a connecting rod to the tiller, the latter being hydraulic pump are the Waterbury type, size 5 A-end, connected by a flexible coupling to an electric motor of 7.1 hp at 350 rpm constant speed. These units are located in the forward and after torpedo rooms respectively.

mounted on and secured by clamping and doweling to the hexagonal-shaped end of each plane stock. The planes are actuated between stops, allowing a total travel of 54 degrees of 27 degrees each side of zero tilt.

14B2. Stern planes. The horizontal athwartship for the stern diving planes is keyed to a tiller which is operated through an angle of 54 degrees by a connecting rod, cylinder, and piston arrangement located in the after torpedo room under the non-watertight walking flat.

The bow and stern diving gears are of the electrohydraulic type and under normal operation the planes are tilted by power. The

The motor of the electrohydraulic system is started and stopped by a push button type switch, and provided with an electrically operated brake which grips the armature shaft when the motor is not being operated. An interlock switch, controlled by operation of the change valve at the operating station in the control room, prevents starting the motor except when the change valve is in *Power* position.

In the control zoom at the operating station, a handwheel is connected to another hydraulic pump. (See Figure 14-1.) Rotating the handwheel with the change valve set for *Power* forces oil to control cylinders at the electrically driven pump. This actuates the

control shaft on the pump by means of a rocker shaft and bell crank, thereby controlling the direction of flow and quantity of oil. Thus, the direction and amount of tilt are governed by the rotation of the handwheel of the diving station; to the right for

Dive and to the left for *Rise*. The handwheel should always be brought back to its original position to stop further movement of planes; this also allows the centering device, consisting of a double-acting spring, to hold the pump control shaft in its neutral position.

C. OPERATIONS

14C1. Power tilting of planes. When tilting the bow or stern planes by power, the change valve for that plane must be in the power position. The stroke adjuster is set for one-quarter of a stroke and the hydraulic pump motor is started at the diving station. When the station is set for power tilting, turning the wheel delivers oil to the control cylinder which actuates the tilting box of the power-driven hydraulic pump. Tilting the box of a power-driven hydraulic pump causes it

similarly, they are not interconnected and must be individually operated.

Turning the emergency control valve handwheel clockwise directs oil from the main hydraulic system directly to the stern plane ram in the after torpedo room and the bow plane rain in the forward torpedo room. The oil flow will be for dive. Turning the control valve handwheel counterclockwise admits oil to the stern plane ram and the bow plane to deliver oil to the tilting ram.

When the wheel is turned clockwise, oil is delivered for dive angle on the planes; when the wheel is turned counterclockwise, oil is delivered for rise angle on the planes. (See Figure A-22.)

14C2. Hand tilting of planes. In tilting the bow and stern planes by hand, the electrohydraulic system is stopped and the change valve at the diving station is set for *Hand*; rotating the handwheel forces oil directly to the hydraulic cylinder located in the forward torpedo room for bow planes, and the after torpedo room for stern planes.

The bow plane setup is the same as the stern plane setup for hand tilting except that the *hand rigging* and *tilting control* valve is used on bow plane tilting to direct the oil to the rams (Figure A-23) when set in the tilt position.

14C3. Emergency tilting of planes. Provision is made for tilting the planes with the main hydraulic system by setting the change valve at the diving station in the Emergency position. The direction of oil flow for tilting the planes is then controlled by the emergency control valve.

The bow and stern planes have separate control valve handwheels for emergency operation. While the handwheels operate

ram for rise. (See Figure A-25.)

14C4. Plane angle indicating system. The plane angle indicating system is of the selsyn type. The bow plane transmitter is located in the forward torpedo room and is driven from the starboard plane stock by a gear segment and pinion. The stern plane transmitter is located in the after torpedo room and is driven by an arrangement of levers and links from the plane tilting piston rod. The indicators are mounted on the diving station panel in the control room, In addition, an auxiliary plane angle indicating system is provided which indicates the degree of rise or dive on the planes in 5-degree intervals. A mechanical plane angle indicator is provided in the forward torpedo room and two plane angle indicators in the after torpedo room, one on the after bulkhead and one at the tilting cylinder beneath the platform deck.

14C5. Stern diving gear and capstan motor. The stern plane tilting motor, located in the after torpedo room, is also used to operate the after capstan by a silent chain drive with the sprocket mounted between the motor and the hydraulic pump. The chain is removable and can be dissembled by removing a special pin from the links. The stern plane tilting hydraulic pump cannot be cut out when the capstan is in use; therefore, the pump control shaft should be in the neutral position to prevent the movement of the planes.

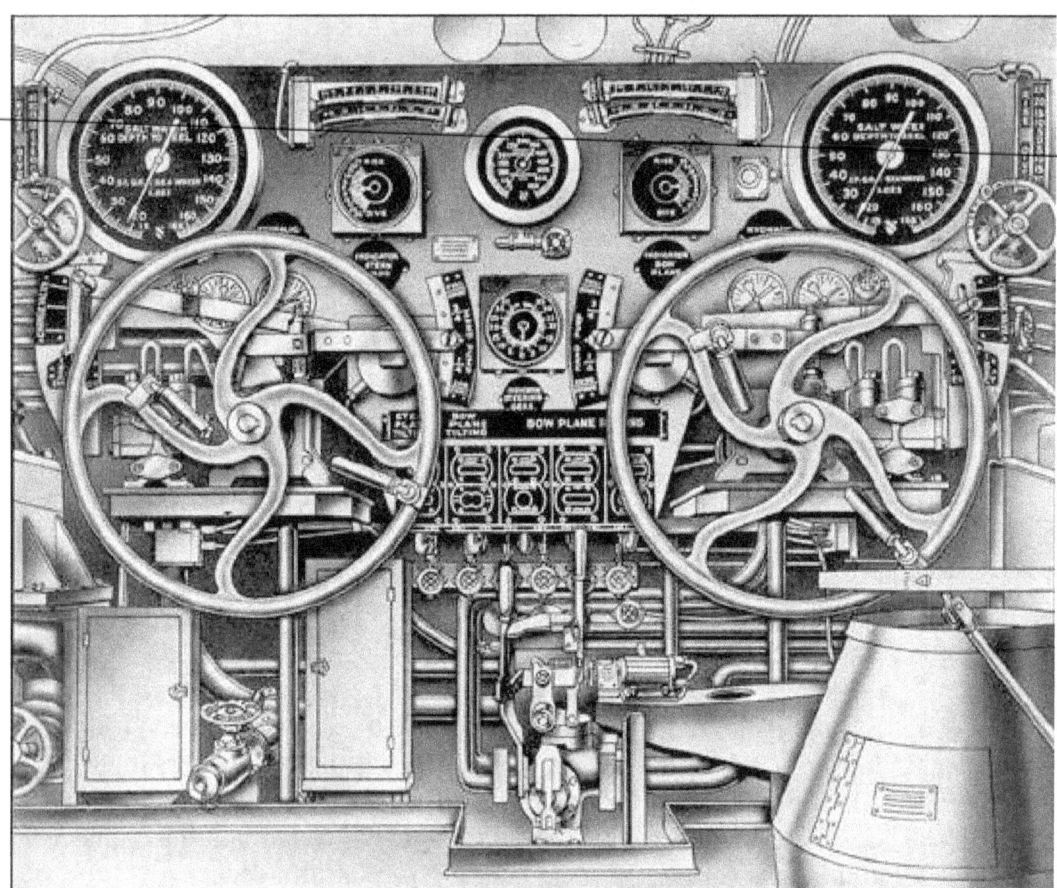

Figure 14-1. Diving control station

14C6. Stern plane drift stop. The position at which zero lift occurs with relation to the stern planes is with the planes set at 4 degrees in the rise, direction; the position at which minimum drag occurs is with the planes set at 2 1/2 degrees in the rise direction. Therefore, it is desirable that the indicator in the control room read zero when the planes are set at 4 degrees in the rise direction, because in that position the lift, when operating submerges, is zero. It is also desirable that when operating on the surface, the planes be set at that angle at which a minimum of drag or resistance is encountered. In order to obtain a setting of 2 1/2 degrees rise while

control valve to rig. (See Figure A-24.) Rotating the bow diving handwheel to the right will rig out the planes. A hand-operated clutch is provided for windlass and capstan or plane rigging operation and is located in the forward torpedo room. With the clutch set for bow rigging operation, the hydraulic motor drives a vertical shaft through a reduction gear and worm gear arrangement. This shaft passes through the hull to the outside where bevel gears and shafting operate two pairs of 39 1/4-inch diameter spur gears. These gears are mounted in pairs athwartships and fore and aft, so that the forward and after gear of each pair carry at a point on the circumference the pin for the crank operating on the surface, the indicator in the control room should read 1 1/2-degree dive.

When cruising on the surface under normal conditions, a drift stop is provided to prevent the stern planes from passing beyond the 2 1/2, degree rise position. This drift stop is located at about the midstroke of the end of the piston rod extension in the after torpedo room. A bar is provided to be lifted in the path of the end of the piston rod extension when the planes are approaching the 2 1/2-degree rise position blocking the path and thus preventing the rod extension from moving farther in the rise direction. The planes must be moved in the dive direction until the indicator (in the control room) reads at least 5 degrees dive before the stop bar can be lifted into position. A pin in the rod extension enters a hole in the bar as the planes are moved to the 2 1/2-degree rise position. The stop bar is released by moving the planes in the dive direction until the indicator (in the control room) reads at least 5 degrees dive, disengaging the pin from the stop bar and allowing the bar to drop clear of the path of extension rod.

14C7. Bow plane rigging arrangement. The bow planes are rigged out for tilting operation by the windlass and capstan hydraulic motor, a Waterbury size 10 B-end normally driven from the main hydraulic system through the rigging control valve. Provision is also made for rigging the planes by hand from the diving station by setting the change valve to *Hand*, the rigging control valve to *Neutral*, and the hand rigging and tilting

end of the plane rigging connecting rod. The gears are rotated 181 degrees 41" between mechanical stops when raising the planes from the horizontal to the housed position. At the lower end of the vertical shaft, an indicator switch is driven by bevel gears; this indicates the *rigged in* or *rigged out* position by telltale lights at the diving station.

14C8. Rigging and tilting interlocks. At the lower end of this same vertical shaft, the rigging interlock switch, driven by bevel gears, prevents the operation of the electrohydraulic bow plane tilting motor except when the planes are in the fully rigged-out position. There is also a hydraulic interlock valve which prevents the flow of oil to the plane tilting cylinder from the hand or emergency control valve until the planes are in the fully rigged-out position. The operating mechanism for this interlock valve is also driven by bevel gears from the vertical shaft, and the bow plane rigging indicator is mounted on the underside of the casing. If the sheer pin drive at the lower end of the pointer shaft fails, the interlock valve may be operated by applying the T-handled socket wrench to the squared lower end of the shaft extending through the pointer hub. This wrench is stowed on the underside of the worm gear unit adjacent to the valve operating mechanism.

In addition, there is also a regulating control valve operated from the same shaft. The purpose of this valve is to restrict the flow of oil to the hydraulic motor when the bow planes are nearly rigged out or nearly housed. The flow is not completely cut off; a small amount is bypassed to permit the planes to creep into the stops located on the large spur gears. There is also a lock on the rigging control valve operated by a solenoid; this prevents the housing of the bow planes unless the planes are within 1 1/2 degrees either side of zero tilt, and then only by manually operating the push button which releases the lock. An additional release of the solenoid lock is also provided for emergency use and is accomplished by means of a lever located in the armature end of the solenoid. This means of disengaging the solenoid lock should be used for rigging out planes only in case they have been thrown off the position for rigging as a result of wave slap.

The diving gear control stations, bow and stern, are located together on the port side of the control room. (See Figure 14-1.) They are provided with an 8 1/2-inch depth gage reading to 450 feet and a rudder angle indicator. Each station is provided with an electrical self-illuminated selsyn type plane angle indicator, an auxiliary plane angle indicator reading at 5-degree intervals, a 16-inch depth gage reading to 165 feet, and a pair of spirit trim indicator inclinometers for angles from 0 degrees to 5 degrees and from 0 degrees to 15 degrees.

15
ANCHOR HANDLING GEAR AND CAPSTANS

A. ANCHOR HANDLING GEAR

15A1. General. The ground tackle consists of one 2,200-pound stockless anchor and 105 fathoms of 1-inch die-lock steel chain. The anchor is housed in the hawsepipe in the superstructure. The anchor chain is self-stowing in the chain locker.

The windlass consists of a wildcat driven by a hydraulic motor, which also drives the forward capstan (Figure A-26). The motor is a Waterbury size 10 B-end, and is operated from the main fair-lead upon which are mounted two jaws operated by two traveling nuts on a screw to act as a chain stopper for securing the chain. The chain stopper is operated by gears and shafting from the main deck. An indicator is provided at this station and is operated from the drive gear casing for the chain stopper operating shaft.

The bitter end of the anchor chain is secured in a tumbling hook just above the top of the chain locker and is unlocked

hydraulic system.

The hydraulic motor drives a worm gear, located overhead, which in turn drives a vertical shaft passing through the hull.

The windlass and capstan clutch, located in the drive gear casing at the upper end of the vertical shaft, is operated from the main deck, an indicator being fitted at this station. A pair of spur gears in the drive gear casing drive the vertical shaft on which the capstan head is mounted at the main deck. A pair of bevel gears drive the shaft, which goes forward to a bevel gear assembly. This assembly is mounted on the windlass which actuates the wildcat drive shaft, from which a spur pinion drives the spur gear on the wildcat casting.

A chain indicator, reading in fathoms, is provided on the main deck. Windlass and capstan cannot be driven at the same time.

The wildcat carries a snubber brake band which is operated by gears and a line of shafting to the main deck. An indicator is fitted at this station.

The anchor chain, upon leaving the chain locker, passes through a closed from the main deck. The latch, which releases the hook, is attached by means of a long connecting rod to a crank on the operating shaft, located beneath a small hatch in the deck on the port side, about 15 inches from the centerline of the ship, where a wrench provided for operating the chain slip is stowed in a convenient place. The operating shaft is locked in the normal secured position of the tumbling hook by a hinged locking plate which engages the square end of the operating shaft. The locking plate is held in position by a toggle pin. The anchor chain can be slipped only when the vessel is on the surface.

Hand operation of the windlass from the diving station is accomplished by setting the change valve to HAND, and the rigging control valve to NEUTRAL, the hand rigging and tilting control valve to RIG, and the clutch in the forward torpedo room in windlass and capstan position. Rotating the bow diving handwheel to the right will hoist the anchor.

B. CAPSTANS

15B1. General. Two capstans are provided on the main deck, one between frames 20 and 21 and the other between frames 117 and 118. Both are permanent fixtures mounted on vertical shafts. The capstan heads are 15 inches in diameter for a rope speed of 60 feet per minute.

The forward capstan is driven by a hydraulic motor. (See FigureA-26.) The capstan head is removable and when not in use can be stowed below the main deck on the port side adjacent to the capstan shaft.

The after capstan is driven by the stern plane tilting motor through a chain drive from a sprocket on the after end of the motor shaft to the capstan worm gear. The chain is removable and can be disassembled by removing a special pin from the links. The stern plane tilting hydraulic pump cannot be cut when the capstan is in use; therefore, the pump control shaft should be in the *neutral* position to prevent movement of the planes. The capstan is mounted on the top of the vertical worm gear shaft.

C. OPERATIONS

15C1. Dropping the anchor from the water's edge. Following is the procedure for dropping the anchor from the water's edge:

1. Shift the clutch in the forward torpedo room to the *windless and capstan position.*
2. Shift the clutch on the main deck to the *windless* position.
3. Open the chain stopper.
4. Loosen the brake band.
5. Operate the windlass and capstan control valve from the main deck by turning counterclockwise to lower.
6. When the anchor reaches the water's edge, return the windlass and capstan and anchor control valve to the *neutral* position.
7. Tighten the brake band.
8. Put the windlass and capstan clutch on the main deck in the *neutral* position.
9. Release the brake band.

Note. When the brake band is released, the weight of the anchor will carry out the chain. The chain is snubbed when the desired length is run out.

15C2. Hoisting the anchor by hand. Hoisting the anchor by hand is accomplished as follows:

1. Shift the change valve in the control room at the bow plane diving station to the *hand* position.
2. Set the bow plane tilting stand pump stroke to *full stroke.*
3. Set the hand rigging and tilting control valve to the *rig* position.
4. Shift the clutch in the forward torpedo room to the *windlass and capstan* position.
5. Shift the clutch on the main deck to the *windlass* position.
6. Loosen the brake band.
7. Open the chain stopper.
8. Rotate the bow plane wheel clockwise to raise the anchor.

15C3. Securing the anchor for sea. In securing the anchor for sea, it is necessary to:

1. Shut the chain stopper.
2. Tighten the brake band.
3. Shift the clutch on the main deck to the *windless* position.

16
FUEL AND LUBRICATING OIL SYSTEMS

A. FUEL OIL SYSTEM

16A1. General description. All fuel oil stowage tanks lie between the inner and outer hulls. There are four normal fuel oil tanks No. 1, frames 35-41; No. 2, frames 41-46; No. 6, frames 93-99; No. 7, frames 99-107. There are six reserve fuel oil tanks: No. 3A, frames 57-62 starboard; No. 3B, frames 57-62 port; No. 4A, frames 69-75 starboard; No. 4B, frames 69-75 port; No. 5A, frames 75-80 starboard; and No. 5B, frames 75-80 port. (See Figure A-7.) These six reserve fuel oil tanks make up the fuel ballast tank group.

Each tank is connected to three piping systems: the fuel oil filling and transfer; the compensating water; and the 225-pound service air system. The 3-inch filling and transfer main, with a main deck filling connection at frame 33 and another at frame 69, has two branches to each fuel oil tank, one for each side of the tank. The 3-inch compensating water main, with a deck hose connection at frame 88 and an outlet through a head box, has a branch to the bottom of each fuel oil tank. (See Figure A-27.)

Between frames 91 and 93, there are two tanks: the *expansion tank* on the port side and the *collecting tank* on the starboard side. Located in the conning tower sheers is a head box with a vent and overflow which is kept filled with water from the main engine circulating water system. A 3-inch line carries the water from the head box to the bottom of the expansion tank. Another line is run from the top of the expansion tank

The 3-inch fuel filling and transfer main has a branch to the bottom of the collecting tank, and another line is run from the top of this tank to the engine clean fuel oil tank. Hence, water under pressure from the head box passes through the expansion tank, to the compensating main, and them to the fuel oil tank being used as the supply. From the supply tank, oil flows through the transfer main to the collecting tank and on to the clean fuel oil tank. The stop valve in the collecting tank supply from the 3-inch transfer main is locked open; so locked, valves provide for transfer and compensation under all conditions of operations when a tank, either normal or reserve, is open to the transfer main. A supply tank must always be open to the filling and transfer main, otherwise the collecting tank is subjected to the depth pressure when submerged.

Filling operations are effected through the forward and after deck connections described above. The head of oil forces the water overboard through the deck hose connection for the compensating main or through the expansion tank and head box.

In case of damage to the head box, pressure is kept on the compensating system by a line to the compensating main from the main motor circulating water system in the motor room. Normal operation requires that the stop valve in this line be locked shut. This practice of locking the stop valve during normal operation assures that the compensating system will operate without drawing to the 3-inch compensating water main. These two lines and all branches off the main are provided with stop valves. Each one of these stop valves, except for the hose connection, is kept locked in the open position. The head of water keeps the tanks completely filled at all times. Thus provision is made for the change in volume caused by the variation in temperature and also for changes in pressure, so that the tank is always equalized with sea pressure when submerging.

water from the main motor circulating water system. On the surface, the main engine circulating water system is sufficient to keep the head box full, thus maintaining a constant pressure on the expansion tank and from it to the fuel oil tank on service and the collecting tank. Fuel oil leaving the collecting tank under the compensating system pressure may either go to the purifier and from there by gravity to CFOT, or it stay go to the fuel transfer and purifier pump. The fuel

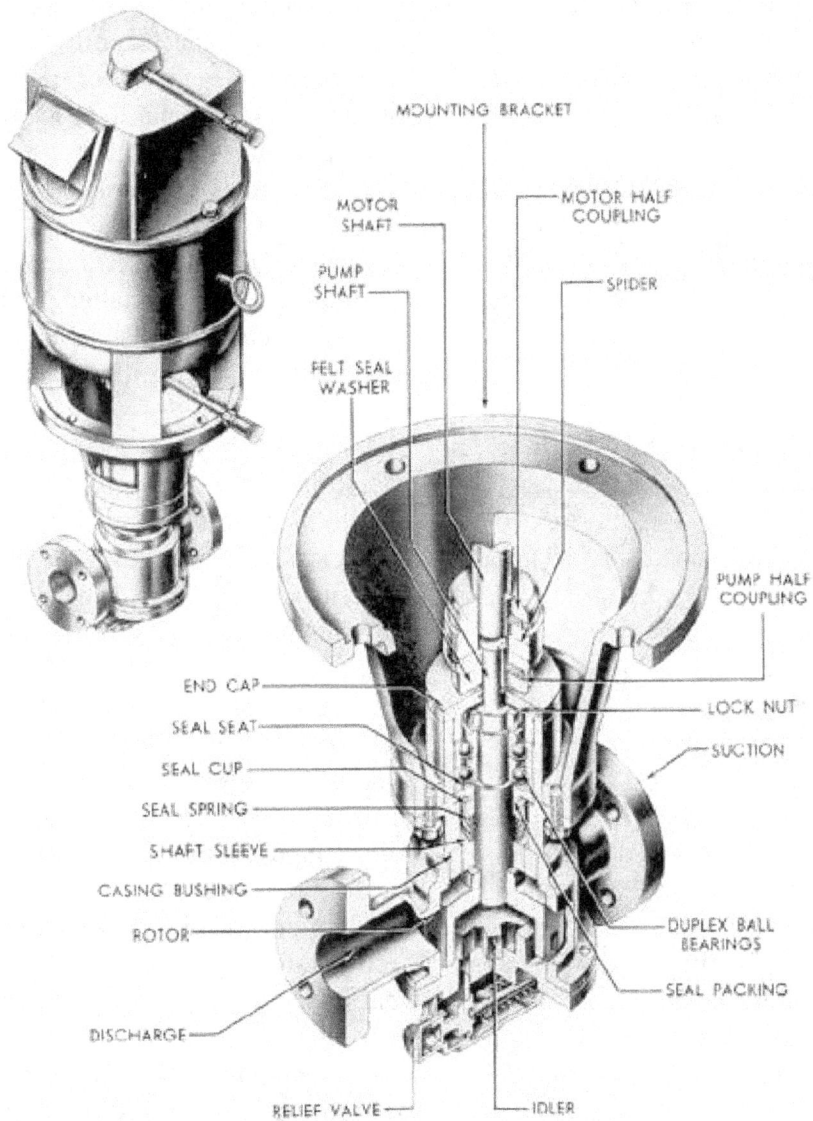

MOUNTING BRACKET

MOTOR SHAFT

PUMP SHAFT

FELT SEAL WASHER

MOTOR HALF COUPLING

SPIDER

PUMP HALF COUPLING

END CAP

SEAL SEAT

SEAL CUP

SEAL SPRING

SHAFT SLEEVE

CASING BUSHING

ROTOR

LOCK NUT

SUCTION

DUPLEX BALL BEARINGS

SEAL PACKING

DISCHARGE

RELIEF VALVE

IDLER

Figure 16-1. Fuel transfer and purifier pump.

164

transfer purifier pump discharges fuel oil to the main engines. The pump may therefore serve as a standby for the main engine fuel pump since the fuel transfer purifier pumps can take a suction on the line from the collecting tank to the purifier; the transfer main and the CFOT discharge this oil directly to the main engines. The fuel transfer purifier pumps may also discharge directly overboard. Purifying drain pump is provided with a discharge to the compensating water main to permit pumping the bilges to the top of the expansion tank, forcing water overboard from the bottom of the tank. This removes the necessity of discharging bilge water directly overboard.

Valves are provided on the branches from the transfer main to the deck filling connections for obtaining samples of the

is not normally done submerged. However, if it is, the depth pressure will maintain a constant pressure on the compensating system.

There are two fuel transfer and purifier pumps: one in the forward engine room, supplying clean fuel oil tank No. 1, and one in the after engine room, supplying clean fuel oil tank No. 2.

The fuel pump is the positive displacement gear type. A d.c. motor of 1 horsepower furnishes the driving power at 1150 rpm operating speed. This pump has a capacity of 10 gallons per minute.

The operating parts include the valve body, the rotor and idler, and the relief valve which is set to open at 45 psi. (See Figure 16-1.)

A bypass line is provided from the transfer main to the fuel transfer and purifier pump for use in the event of bilging the collecting tank. Also, a branch from the head box is run to the compensating main so that the expansion tank may be bypassed. The stop valve in each of these lines is normally locked shut.

The collecting tank is provided with a drain line to the drain pump for the removal of water from the bottom of the tank. The

oil. Fuel tanks are provided with a blow and vent manifold, with blow and vent connections to each side of the tank from the manifold. The manifold is provided with gages and relief valves set at 15 psi. When blowing a fuel tank, water must leave the tank through the compensating water main. Balanced hydraulic gages (liquidometers) are installed in the expansion, collecting, and clean fuel oil tanks to indicate the oil content.

The connection from the compensating. water line to the compensating water manifold and to the head box is provided with a sight glass to check the pipe contents when filling, blowing, and pumping operations are being conducted.

The fuel oil tanks and the compensating water line should be vented frequently to prevent formation of air pockets. Filed oil tanks should be vented in the following order expansion tank, fuel oil tank or tanks in use, collecting tank. The discharge from collecting tank to clean fuel oil tank should be shut when venting.

It is essential that all air be excluded from the fuel oil system, as fuel will not readily flow past an air pocket under the small head pressure provided by the head box.

B. RESERVE FUEL OIL TANKS

16B1. General. Fuel ballast tanks Nos. 3A, 3B, 4A, 4B, 5A, and 5B may be used either as ballast tanks or as reserve fuel oil tanks. When used for reserve fuel oil tanks, the fuel ballast tanks must be isolated from the 600-pound

and the flood valves to these tanks locked shut. This prevents either the compressed air or the sea water from entering the tanks and interfering with the proper operation. of the fuel oil system.

main ballast tank blowing system, and the 10-pound main ballast tank blowing system. All the blow valves leading to these tanks from both the 600-pound and the 10-pound blowing manifolds must be secured,

The main and emergency vent valves to the fuel ballast tanks must also be secured.

Each of the fuel ballast tanks is provided with one vent valve. Blank flanges are provided for the valve openings in the reserve

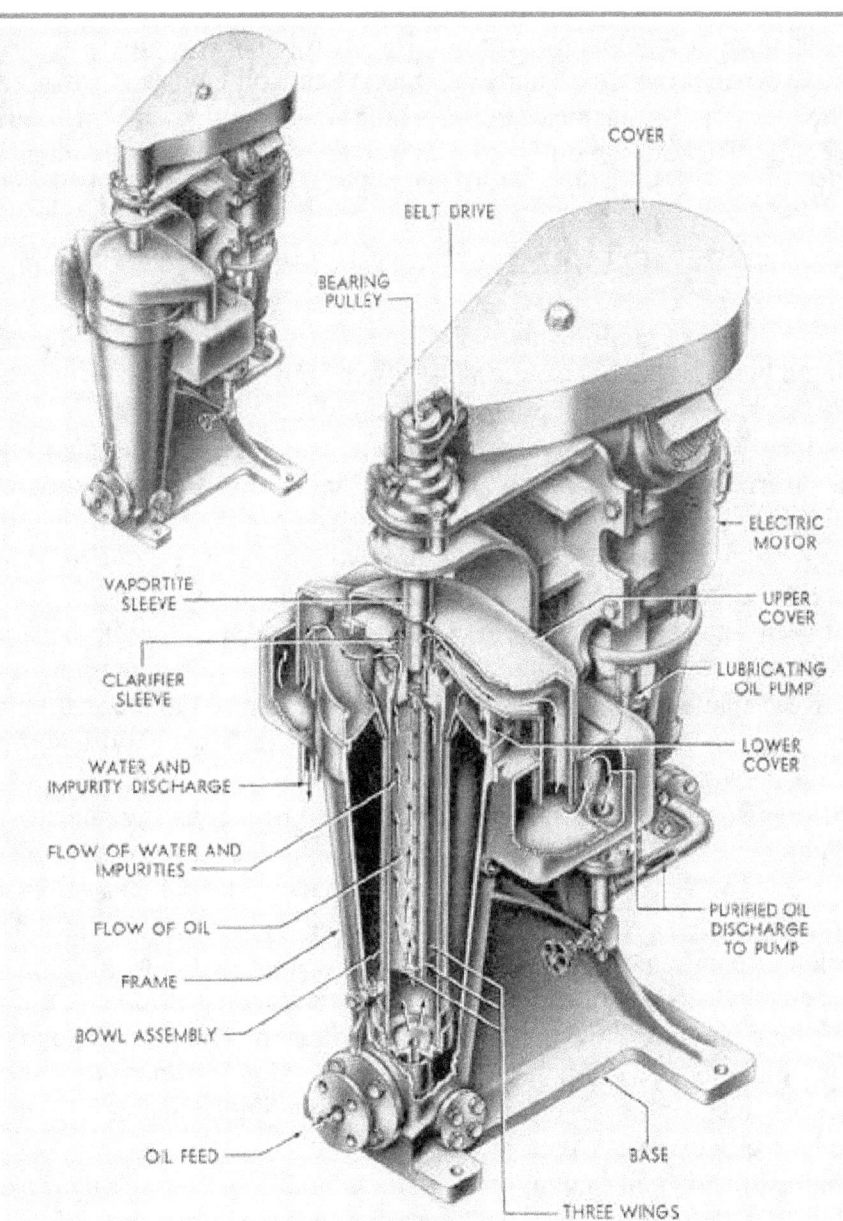

Figure 16-2. Lubricating oil purifier.

fuel oil tanks and are to be installed when fuel is carried in the tank. These blank flanges are stowed in the superstructure, adjacent to the vent valves, and the gasket used with the blank flange is stowed in the ship.

To attach the blank flange, proceed as follows: Open the vent valve a slight amount to relieve the tension on the spring mechanism. Remove nuts, tap bolts, and distance pieces from studs in the finished face on the outside of the valve openings. Remove the pin through the lower end of the bellcrank and connecting link and put the gasket in place over the studs. Attach the blank flanges in place over the studs and replace the nuts and tap bolts and set up to make a tight joint. Put the distance pieces in a bag, label them, and stow in ship. After the blank flanges are in place, return the operating mechanism to the locked shut position.

The emergency vent valves for the reserve fuel oil tanks are provided with padlocks. When fuel is being carried in these tanks, the emergency vents will be locked shut.

C. LUBRICATING OIL SYSTEM

16C1. General. There are three normal lubricating oil tanks: No. 1 between frames 79-85, No. 2 between frames 90-96, and No. 3 between frames 107-109. One reserve lubricating oil tank is located between frames 76-77 port. There are four main engine sump tanks: No. 1 between frames 80-85 starboard, No. 2 between frames 80-85 port, No. 3 between frames 91-96 starboard, and No. 4 between frames 91-96 port. There are two reduction gear sump tanks: No. 1 between frames 103-105 starboard, and No. 2 between frames 103-105 port.

A filling connection is provided on the main deck between frames 78-79 port; it is connected by a 2-inch line to a four-valve lubricating oil filling and transfer manifold located at frames 78-80 starboard. This manifold is connected directly to each of the normal lubricating oil tanks and reserve lubricating oil tanks.

An accessory to the lubricating oil system is the lubricating oil purifying and flushing system. The principal part of this system is the lubricating oil purifier (See Figure 16-2.), used to separate impurities from the lubricating oil. There are two lubricating oil purifiers: one located in the forward engine room amidships and the other in the starboard forward end of the after engine room. The major part of the lubricating oil purifier is a hollow cylindrical rotor called the *bowl*, the top part of which is connected by a coupling nut to a spindle, which in turn is attached to and suspended from a ball bearing assembly. Three flat plate wings are spaced radially equidistant inside the bowl. This three-plate assembly has a cone on the bottom with which the feed jet comes in contact, thus increasing the liquid flow evenly and eliminating emulsion formation.

The spindle is belt-driven by an electric motor fastened on the back of the frame. The belt tension is maintained by an idler

The tanks are normally filled by passing the oil through a strainer before it reaches the manifold; however, this strainer may be bypassed. The lubricating oil tanks are provided with a blow and vent manifold with blow and vent connections from the manifold to each tank. Air is furnished from the 225-pound service air lines through a reducing valve set for 13 psi and a relief valve set for 15 psi. Oil may be blown from any storage tank to any other tank; also, oil to be discharged may be blown or pumped overboard through the deck filling connection or through a 1 1/2-inch hose connection from the filling line.

pulley to assure smooth acceleration.

When the pump is running, the liquid is jetted into the bowl and, upon coming in contact with the three flat plates, is rotated at the speed of the bowl. This rapid rotation causes centrifugal force to act on the liquid, thus separating the heavier from the lighter components in the liquid. Solids, sludge, and water (the heavier parts) are forced through the oil layer to the outside where they form a layer on the wall of the bowl.

As the oil in a purified condition reaches the overflow or discharge port, it is carried

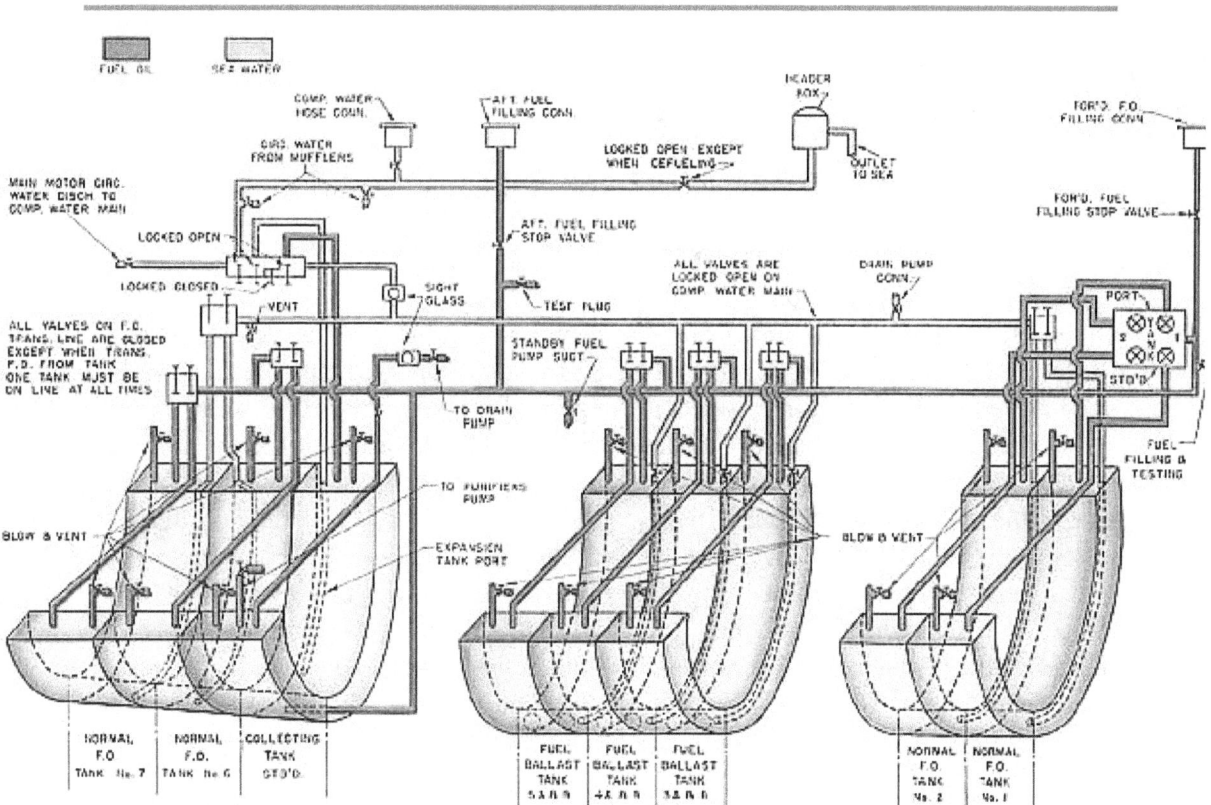

Figure 16-3. Fuel oil system.

to the lubricating oil pump and distributed to the engines.

which also serves as the fuel oil purifier just described. The machine is set up as

The lower end of the bowl is set in a guide bushing which shifts in accordance with the position required by the center rotation of the bowl.

The lubricating oil purifying and flushing system may be used to perform the following functions:

1. Flush engines prior to starting.
2. Purify the lubricating oil.
3. Supply engines with lubricating oil.
4. Clarify the lubricating oil.

Clarification of the lubricating oil is accomplished by the *Sharples centrifuge*

a clarifier by installing a clarifier sleeve or ring dam on the top of the bowl, thus closing the outlet passage through which the water is discharged. The term *clarifier* is applied to the machine when it is set up to discharge a single liquid from which solid matter has been removed by centrifugal force. If the machine is set up to separate two liquids from solid matter and from each other (such as oil and water in a fuel oil purifier), it is called a separator. The machine is usually set up as a separator for fuel oil purification and a clarifier for lubricating oil purification.

169

17
GENERAL SURFACE OPERATIONS

A. OFFICER OF THE DECK

17A1. Duties and responsibilities. Surface operation requires that a submarine be maintained in a constant state of readiness, prepared at all times for any emergency. The safety of the ship is the responsibility of the officer of the deck, subject to any orders he may receive from the commanding officer. The manner in which the officer of the deck should carry out his duties is fully described in *United States Navy Regulations* and the *Watch Officer's Guide*. However, the unusual features of a submarine, which directly affect the officer of the deck, make it imperative that emphasis be given those points that are peculiar to this type of vessel.

securing of all loose gear both topside and below. Superstructure access openings must be locked or welded closed, in anticipation of the most extreme conditions, to prevent accidental release of gear and consequent indication of the submarine's position, and all external air and oil leaks must be eliminated. Bow and stern tubes must be loaded with torpedoes ready to fire and with one complete set prepared for reload without further adjustment. All guns and ammunition must be maintained in the most advanced state of readiness compatible with their preservation in event of quick dives.

17A4. Vulnerability of ship. The

17A2. Desirable characteristics.
Desirable qualities of a submarine's officer of the deck are forehandedness, vigilance, leadership, and common sense. Each of these qualities must be developed to a high degree of perfection to insure the successful execution of duties. The officer of the deck's ability to look ahead and foresee the development of unusual contingencies places him in a position of always being prepared. On submarines, more so than on other types of vessels, it is axiomatic that eternal vigilance is the price of safety. Alert to all that is going on about him, he must see everything and know all that is happening. Conducting himself in a manner befitting his position, as representing the commanding officer, requires a marked degree of leadership. To this extent he must be exemplary in his appearance and performance of duty so that a feeling of confidence and pride is developed in his subordinates.

The underlying attribute of all these characteristics of the proficient submarine officer of the deck is common sense-a sense of proportion, the ability to evaluate the components of a situation in the light of their real significance.

17A3. Preparing the ship. Preparation must be completed before leaving port to place the submarine in a rigged-for-dive condition. The officer of the deck supervises the stowing and

officer- of the deck must have a more intimate technical knowledge of all departments of a submarine than is required in surface ships, because a submarine on the surface is the most vulnerable type of craft afloat. The danger of collision exists on a submarine long before that condition is considered to exist on a surface vessel; therefore, the safety of the submarine must never be jeopardized by unnecessarily placing it in a position in which collision with another vessel is possible. Low reserve buoyancy magnifies the danger resulting from any collision, requiring the submarine to be kept well clear of all vessels. Things happen quickly in submarines, requiring quick thinking and decisive action to grasp the opportunity to prevent disaster. Taking halfway measures may lose the ship.

17A5. Watch officer's station. The forward bridge structure is the usual station of the officer of the deck when on the surface. Although he is expected to remain intensely alert and observant, he is not a lookout, and must not become engrossed in details to the exclusion of his comprehensive duties as supervisor of the watch. The poor habitability of the bridge and the exacting requirements of the duties impose strict demands upon the assigned personnel. The officer about to relieve the deck should be properly clothed and physically ft to assume the responsibilities

that will be his. For similar reasons, lookouts should be selected from the best men of the crew suitable for this

He must insist that acknowledgement of directives be made in standard phraseology, permitting no deviations

duty, and only those chosen who have excellent vision and good health. Prior to their coming on the bridge, clothing should be issued to enable them to withstand the rigors of adverse climatic conditions, Cold, wet personnel cannot function as efficiently as those who are protected from the weather, in so far as conditions will permit. The night lookout should be properly dark-adapted, wearing red goggles, before relieving.

17A6. Conduct of the watch. The officer of the deck must take over his watch promptly, and be sure to obtain accurately all the information from the officer about to be relieved.

He should determine the ship's position with regard to other ships in sight, the proximity of land, rocks, shoals, and the identity of lights. He should always time navigational lights used in fixing the ship's position even though they have been previously sighted and identified. He must keep the ship's position plotted on the chart at all times. He must closely observe the weather, course, speed, and know the combinations of propulsion equipment available. He must know whether or not the storage batteries are being charged and how much float is being carried. He should insure that battery ventilation is adequate.

He should maintain an efficient watch by rotating the lookouts. He should stagger the reliefs, allowing only one man of the oncoming watch on the bridge at a time and he should require "permission to come on the bridge" in each case, and caution against all extraneous noise and unnecessary conversation. Orders should be worded whatsoever.

The success of night attacks depends greatly upon the alertness and reliability of the lookouts. Each should be trained to know what to look for, carefully searching his assigned sector, and reporting his findings in the proper phraseology. When an unidentified or enemy vessel is sighted, the officer of the deck should be so familiar with the commanding officer's attack doctrine that he can take the proper action while calling the crew to battle stations. In submitting reports to the commanding officer, he must be certain that the data are correct, with any doubtful details so identified. He should develop a reputation for reliability and integrity. He must be sufficiently familiar with signals to be able to determine when another ship is calling without having to call for the quartermaster every time flashing lights are seen.

The officer of the deck must have full information of the status of every department of the ship at all times. He must have knowledge of the condition of all hull openings, ballast tanks, flood valves, vents, variable tanks, pumps, and so forth. Particular attention must be given to the ship's readiness to dive, permitting nothing to jeopardize this condition without the commanding officer's permission. As soon as charts, sextants, and other loose gear are no longer in use, or the necessity of additional personnel on the bridge has ceased to exist, he should see that the ship is returned to a condition in which she is able to dive without delay. Appropriate consideration should be given the fact that submarines have a small freeboard, resulting in danger of personnel being washed overboard or

in standard phraseology, and given in an authoritative manner only as loud as the occasion demands.

water entering the ship through ventilation and deck openings.

B. SHIP HANDLING

17B1. Experience. The mere reading of a book will not establish perfection in the art of ship handling, an accomplishment attained only by practice, and more practice, in performing the actual operations. Nevertheless. the printed page provides a means of perpetuating the findings and advice of those who have learned by experience.

Above all, ship handling demands good judgment. Existing and anticipated situations must be carefully considered before action is taken. The officer of the deck should

handle the ship smartly, and he should always remember that in coming alongside a dock or another ship, a submarine holds her way longer than a surface ship of similar tonnage.

17B2. Control. Steering and engine control are, at all times, from the conning tower or control room, in which are located the power, emergency, and hand steering controls. The officer of the deck should practice the use of the two latter systems, except when maneuvering in restricted water's, thereby training. the steersmen and testing the equipment for emergency operation. In a submarine, more so than in other types of vessels, good judgment demands that all machinery be tested prior to its prospective use. Hand steering trains steersmen to use small amounts of rudder, and permits more nearly silent operation.

Control is temporarily poor when shifting from conning tower to control room, and when shifting from engines to motors. If a doubtful situation exists at these times, prudence dictates

their direction of rotation should be thought out carefully. Quite often the backing of both propellers would spoil an otherwise good landing, where the situation demanded that one, or the other, should be backed.

17B4. Current effect. Accurate estimation of the strength and direction of current is essential to the success of the maneuver. This is easily estimated by noticing a spar buoy, or watching the water flow past the end of a dock. The current must not only be considered when selecting the turning point while in the channel but it must be remembered after the turn, realizing that the slower the vessel approaches the dock, the greater will be the effect of the current. Similarly, it is important to remember that after the bow enters the slip, it is in relatively still water, while the current continues to produce its full effect on the stern.

17B5. Wind effect. A submarine making sternboard will back into the wind, because of the greater freeboard of the bow, making it difficult to turn in a

maintaining the status quo until circumstances permit a change, When operating on the surface, enough way should be kept on the ship to permit maneuvering or quick diving.

17B3. Landings. When maneuvering around docks and other close quarters, especially at night, the officer of the deck must assure himself of unobstructed visibility in all directions. He should carefully plan the approach to the landing, with special reference to current, wind, amount of way on, turning points, sea room available in the slip, and the preparation of lines. The use of excessive speed is both dangerous and inexcusable; a submarine is not equipped with four-wheel brakes.

He should never bump any part of the submarine-the underwater bow and stern parts are especially vulnerable, Landings should be made gently. Landing should be a precisely executed maneuver, so planned and performed that the simple operation of backing one or both propellers for a few seconds, when near the desired position, will take all, or nearly all, way off the ship and leave her practically in her berth, ready to double up all lines. The choice of screws and

narrow channel or maneuver alongside a dock. Like all propeller-driven surface ships, a submarine rides more easily with her quarter to the wind and Sea .

Possible hull distortion, and damage to bow tube shutters, diving planes, superstructure, and bridge strongly indicate that the submarine should not be pounded into heavy seas unless absolutely necessary.

17B6. Turning. In making a turn in a narrow channel, the ship may be turned on her heel by going ahead slowly on one propeller while backing full on the other, with rudder over in the direction of turn to assist the ahead screw. This forward and reverse combination of the screws may also be helpful in getting the stern in to the dock, Before backing on one or both propellers, the rudder should be used to steady, or start, a desired swing, as conditions may warrant. If the vessel is swinging, the backing of the inboard screw will normally accelerate the swing.

17B7. Backing. Special signals and exact procedure for backing must be established for emergency use in the event of failure of

usual engine signals. Before backing, the officer of the deck should see that all is clear, and guard the stern planes and propeller while proceeding out of the berth. The stern should be placed well clear of the dock by holding a forward spring, while going ahead for a few seconds on the outboard screw. A strong dock will permit winding around

intelligently. The men at the lines should not be expected to do the thinking for the officer of the deck. These mooring operations should be directed just as actively and positively as signaling the engines. One satisfactory method of getting the ship alongside the dock against the tide is by securing a breast line from the bow to the dock. Then, with

the end, after the stern is clear.

17B8. Handling lines. In approaching the dock, the mooring lines should be handled

the rudder outboard, the outboard propeller should be backed while going ahead on the inboard screw.

C. STANDARD PHRASEOLOGY

17C1. Getting underway.

a. "Station the maneuvering watch." Personnel man their stations in accordance with the Watch, Quarter, and Station bill. Start and test machinery. Special details such as line handlers, anchor detail, color detail, and leadsman take their stations.

b. "Stand by to answer bells." A preparatory command to the watch, indicating that orders to the engines will follow directly.

c. "Station the regular sea detail." An order given when clear of restricted waters and the special details of the maneuvering watch are no longer required.

17C2. Line handling.

a. "Stand by the lines." Man the lines, ready to cast off or get the lines over to the dock.

b. "Cast off number one." Release number one line from the dock.

c. "Ease four." Pay out enough of the designated line to remove most of the strain.

d. "Hold three." Take enough turns so that the designated line will not give.

e. "Check two." Hold, but let it run when necessary so that it will not part.

f, "Take a strain on four." Put the line under tension.

g. "Get over number one." Heave number one line to the dock.

h. "Take three up the dock." Man ashore receiving the line takes it up the dock to a new position.

lines." Pull the lines, released from the dock, aboard.

k. "Single up." Bring up double lines so that only single parts remain secured.

l. "Double up and secure." Run additional lines and double them as necessary to secure the mooring.

m. "Slack one (two)." Pay out the line, allowing it to form an easy bight.

17C3. Orders to the wheel.

a. "Right (Left) rudder." A command to give her right (or left) rudder instantly, an indeterminate amount. In all such cases, the officer conning the ship should accompany the order with a statement of his motive, or the object to be attained, so that the steersman may execute the order with intelligence and judgment,

b. "Right (Left) full rudder." A maximum rudder angle of about 35 degrees is used in the Navy.

c. "Right (Left) standard rudder." Not used on submarines.

d. "Right (Left) standard half rudder." Not used on submarines.

e. "Right (Left), 5 (10, etc.) degrees rudder." These orders are used in making changes of course. All courses given to the steersman must be compass courses.

f. "Right (Left), handsomely." This order is given when a very slight change of course is desired.

g. "Give her more rudder." Increase the rudder angle already on to make her turn more rapidly.

h. "Ease the rudder." Decrease the rudder angle already on when she is

i. "Take in the slack on four." Heave turning
on line and hold it taut, but do not take
a strain.

j. "Take in two." or "Take in the
after

too rapidly, or is coming to the heading desired. The order can be given, "Ease to 15 (10, 5, etc.)."

i. "Rudder amidships." Rudder is centered and kept there until the next order.

j, "Meet her." Use rudder as may be necessary to check, but not entirely stop her swing.

k. "Steady," or "Steady so," or "Steady as you go." Steer the course on which the ship is heading when the command is received.

l. "Shift the rudder." Change from right to left rudder, or vice versa.

m. "Mind your rudder." A warning to the quartermaster (or steersman) 1) to exact more careful steering, or 2) to put him on the alert for the next command to the wheel.

n. "Mind your right (left) rudder." A warning that the ship shows a frequent tendency to get off her course, and that if right (or left) rudder is not applied from time to time to counteract this tendency, the ship will not make good the course set.

o. "Nothing to the right (left)." Given when the course to be made good is a shade off the compass card mark, and therefore meaning that all small variations from the course in steering must be kept, for example to the southward of the course set.

p. "Keep her so." A command to the steersman when he reports her heading, and it is desired to steady her.

q. "Very well." Given to the steersman, after a report by him, to let him know that the situation is understood. The expression "All right" should not be used, it might be taken as an order to the wheel.

17C4. Orders for the engines. Standard orders to the engines are given in three parts: 1) the first part designates the engine *starboard, port, or all;* 2) the second part indicates the direction: *ahead or back;* and 3) the third part indicates the speed: 1/3, 2/3, *standard, full, flank, or stop.*

Typical orders are:
1. "Port, ahead, 2/3."
2. "Starboard, back, full."
3. "All, ahead, standard."
4. "Port, back, 1/3; Starboard, ahead, 2/3."
5. "All, stop."

In the submarine service, the word *engine* is omitted in orders to the engines or motors, to eliminate confusion resulting from the fact that under various conditions, and with various types of main drive, engines sometimes deliver the power, and motors sometimes do.

18
GENERAL DIVING OPERATIONS

A. SUBMERGING

18A1. General. A submarine, when on the surface, must be in readiness to dive at any time. This requires the vessel to be maintained constantly in *diving trim* so that little or no trim adjustment is necessary after submergence. The ship is rigged for dive immediately after getting underway, in strict. conformance with the check-off list posted in each compartment.

18A2. Types of dives . Three types of dives are possible: 1) the *quick dive*, 2) the *running dive*, and 3) the *stationary dive*, the choice being dictated by existing conditions. In all dives the ship is placed in the condition of neutral or negative buoyancy; the use of negative buoyancy shortens the diving time.

A *quick dive* is made when the ship is underway on one or more main engines. The bow planers are placed on FULL DIVE and the forward speed results in a maximum downward thrust on the bow planes. As the submarine submerges, the upper surfaces of the hull and superstructure act as planing surfaces and increase the downward thrust. The quick dive is the fastest of the three types and is used in acceptance trials of new submarines, when it must be executed within 60 seconds from *standard diving trim*.

A *running dive* is performed in the same manner in a minimum time, the only difference being that surface propulsion, at the beginning of the dive, is by battery-powered main motors.

18A3. Control station. The necessary instruments and equipment for submergence and control of the boat while submerged are located in the control room. The control room is the station of the diving officer who issues the necessary orders during a dive and directs the men at the manifolds, pumps, and diving gear in maintaining the submarine at the desired depth. An illustration of the control station is shown in Figure 14-1.

The control station includes the bow and stern planesmen's stations and is equipped as follows:

a. *Depth gage*, reading from 0 to 165 feet, indicates the keel depth,

b. *Inclinometers* indicate the angle of the vessel's longitudinal axis with the horizontal.

c. The *plane angle indicators* are miniature diving planes, the trailing edges of which range over degree scales and indicate the angle of rise or dive on the bow and stern planes.

d. A *motor control contact-type switch*. provides selection of direction of rotation of plane-tilting motor in those installations using electrically operated planes. Hydraulically operated planes are controlled by a *two-way valve*.

e. The *selector switch* provides a choice of motors on which the controller is to be used. The bow planes may be rigged or tilted according to the position of the

A *stationary dive* is made when the ship is dead in the water. The main ballast tanks are completely flooded and enough water is flooded into the variable ballast tanks to destroy the remaining positive buoyancy. Flooding and pumping may be alternated to maintain the ship on an even keel at any desired depth. This type of dive is sometimes used during builder's trials to test and inspect the hull at various submerged depths down to final test depth.

switch handle. A similar device at the stern plane controller provides for use at the stern plane motor to operate the after capstan.

f. The *hand-operated clutch*, on electric drive, permits shifting from power to brand operation of planes. Hydraulically operated planes are controlled by a *valve* which, when placed in the neutral position, permits manual drive direct to the planes.

g. A *depth gage* reading from 0 to 600 feet, mounted at the center of the control panel, is visible both to planesmen and the diving officer.

h. The *rudder angle indicator*, graduated in degrees, indicates the amount of left or right rudder on the ship.

i. The *aneroid barometer* indicates the air pressure in the ship.

j. The *hull opening indicator light panel*, commonly referred to as the *Christmas tree* (Figure 18-1), indicates the condition of hull openings, vents and flood valves. Electrical contacts at these various openings and valves operate red and green lights in the panel. A green board shows that all hull openings are closed.

18A4. Rig for dive . Before an actual dive is undertaken, the submarine must be fully prepared. Compensation must have been made for all changes in weight since the last dive, and all hull openings, except those necessary for the operation of the ship on the surface, must be closed. The hull openings not

c. The *main engine induction outboard valve* which admits air to the engines.

d. The *forward and after engine room induction hull flappers* which admit air to both forward and after engine rooms.

The safety and negative tanks are flooded and the flood valves left open. All bulkhead watertight doors and bulkhead flappers are checked for free operation. Outboard battery ventilation exhaust valves, if installed, are closed and checked by sighting the valve disc and by checking the mechanical indicator. Sufficient high pressure air, at least 45 percent of total capacity, must be cut into the manifold. Rapid and efficient communication must be established between the diving officer and all compartments.

All of these details must be checked by one or more officers. When the diving officer is informed that all preparations are completed, he reports to the

closed, and the reasons therefore, are as follows:

a. *Conning tower hatch* which provides passage to or from the bridge.

b. The *hull ventilation supply valve* which remains open to permit the circulation of fresh air.

commanding officer, "Ship rigged for dive."

18A5. Diving procedure . Each ship has a diving procedure which has been established by careful consideration of its particular characteristics, and is recorded in the Ship's Organization Book of that submarine.

The diving signal is two short blasts on the diving alarm, the second blast being the signal to start the dive. Two blasts are used

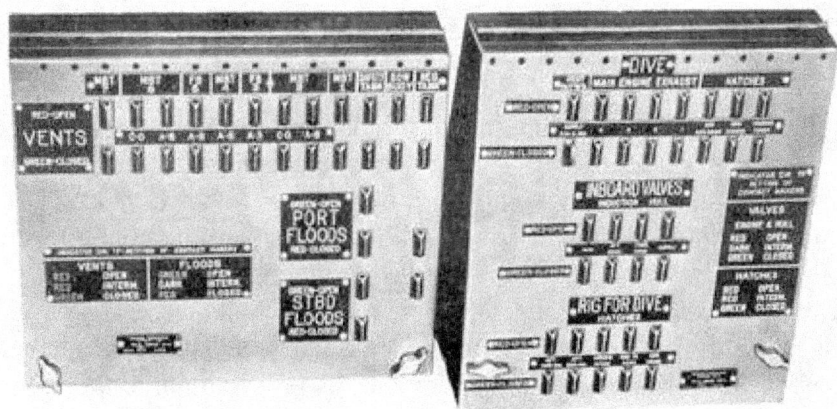

Figure 18-1. The hull opening indicator light panel.

176

to guard against diving on an accidental single blast. An alternate diving signal is the word "Dive, Dive" passed orally.

When the diving alarm is given, the following procedure is observed (items marked with all asterisk are executed at once without further orders)

*a. Stop all engines, shift to battery, set annunciators on "All ahead standard," open engine room doors and air locks.

*b. Close outboard and inboard engine exhaust valves, close hull ventilation supply and exhaust valves, close inboard engine air induction flappers,

*e. Diving officer checks hull opening indicator light panel for condition of hull openings. Air is bled into the ship when green lights show all hull openings closed. Watertight integrity is assured when the internal air pressure remains constant.

f. The following operations are performed by direction of the diving officer, who is guided by the existing conditions:

1. At 45 feet, shut the vents and slow to 2/3 speed.

2. At 15 feet short of desired depth, blow

and close conning tower hatch.

*c. Open bow buoyancy vents and all main ballast tank vents, except the group or tank designated to be kept closed until pressure in the ship indicates that all hull openings have been closed.

*d. Rig out bow planes and place on FULL DIVE. Use stern planes to control the angle on the ship.

negative tank, shut its flood valve, and vent the tank.

3. Level off at specified depth, slow to 1/3 speed, cycle the vents, and adjust fore-and-aft trim and over-all weight.

4. Diving officer reports to conning officer when trim is satisfactory.

When diving in a heavy sea, the maneuver is accomplished most expeditiously by diving with the sea abaft the beam.

B. SUBMERGED OPERATION

18B1. General principles . Submerged operation involves control of the submarine in both the horizontal and vertical planes. Maneuvers in a horizontal plane are controlled by the rudder which functions in the same manner as when on the surface. Control in a vertical plane is achieved by varying the angles of the diving planes and the submarine, combined with an adjustment of ballast to attain the necessary state of buoyancy.

Submerged operations are usually carried on with the submarine in the state of neutral buoyancy. Shortly before reaching the desired depth when diving, the negative tank is blown. This places the ship in the state of neutral buoyancy in which the downward force of gravity is exactly- balanced by the opposite force of buoyancy and the ship will maintain its depth unless acted upon by some unbalancing force. The state of exact neutral buoyancy is probably never attained, but the approximation is near enough to allow depth control to be exercised easily by

speeds. This horizontal motion through the water enables the surfaces of the diving planes to correct the effect of any slight positive or negative buoyancy and also to increase or decrease the submerged depth at the order from the conning officer. The effect of the diving planes is proportional to the speed, and the speed at which the depth can be changed increases with the inclination of the submarine's longitudinal axis from the horizontal. When the axis is inclined, the hull presents planing surfaces. The resultant upward or downward thrust is added to that of the diving planes.

In Chapter 2, *final trim* is defined as the adjustment of ballast while submerged which maintains the submarine at the desired depth, on an even keel, at slow speed, with a minimum use of the diving planes. The diagram in Figure 18-2 represents a typical submarine submerged and in final trim. The forward moments due to gravity, *FGM*, are exactly equal to the opposing moments, *FBM* due to buoyancy. Also the after moments, *AGM* and *ABM*, are equal. To meet the conditions of stable equilibrium,

the diving officer.

the forward

In all normal submerged operations, the submarine is underway at relatively slow

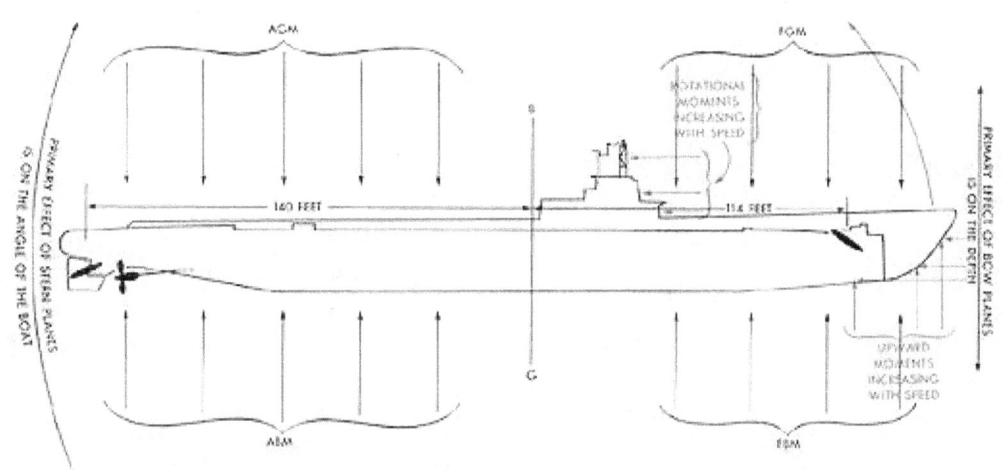

Figure 18-2. Forces affecting depth control.

moments must equal the after moments and the center of gravity must be below the center of buoyancy and in the same vertical line.

When forward motion is imparted to the submarine, new forces, due to the unsymmetrical contour, come into play and disturb the static equilibrium. These forces are indicated in the diagram. The upsweep of the forward end presents a planing surface, the resultant effect increasing with the speed. The superstructure, conning tower fairwater, periscopes, masts, deck guns, and other objects present more or less vertical surface above the center of buoyancy. The tendency of these forces is to rotate the submarine about a center with the result that, as speed is increased, the submarine tends to

is placed in final trim, the diving officer may have control of the speed. It is important, therefore, that the various necessary operations be completed in the minimum time so that speed control may be returned to the conning officer.

18B2. Trim analysis . To place the submarine in final trim the shortest possible time requires a swift, accurate analysis of the submarine's condition. This analysis may be made in two stages: first, the over-all condition is determined and corrected; then, the condition of fore-and-aft equilibrium is determined and the necessary adjustments made. When the fore-and-aft trim is very evident, the necessary corrections should be combined with the correction for over-all trim. When properly made, this analysis conveys a thorough understanding of the

assume a small up-angle and decrease its depth.

This rotational tendency is counteracted by the diving planes. Since the location of the planes is not symmetrical about the center of buoyancy nor about the longitudinal axis of the submarine, the effect of the bow planes on the movement of the vessel is quite different from that of the stern planes.

The location of the planes is shown in Figure 18-2. (The dimensions shown are approximate, and apply only to the submarine described herein.) As indicated, the bow planes tend to change the vertical position of the submarine on an even keel. There is a certain rotational moment, but it is counteracted to a great extent by the longer after body which acts somewhat as a stabilizing rudder, resisting angular displacement.

The stern planes are located at a greater distance from the center of buoyancy and, although they are smaller in area, their effect is much greater than that of the bow planes. Their effectiveness is increased by their location directly aft of the propellers. This combination of a long moment arm and location in the propeller wash results in a rotational movement. Thus the bow planes are normally used to control the depth, and the stern planes the angle of the ship.

From the time the dive is started until the submarine reaches the specified depth and

submarine's condition and indicates the corrective measures.

Certain terms have been established as standard phraseology in describing the trim of the submarine. These terms are factual statements of conditions and also indicate the procedure necessary to attain *final trim*.

a. *Heavy over-all and heavy aft*. The entire phrase is necessary to describe conditions. *Heavy over-all* indicates negative buoyancy and heavy aft indicates that the fore-and-aft moments are not in equilibrium and that there is too much weight aft. The remedy and order would be, "Pump from after trim to sea."

h. *Light over-all and light forward*. Light over-all indicates that the submarine has positive buoyancy and light forward indicates the remedy: "Flood forward trim from sea."

c. *All right over-all and heavy aft*. This indicates that the ship is in the condition of neutral buoyancy and that equilibrium should be established fore and aft by the adjustment of ballast between the trim and auxiliary tanks. This adjustment may be obtained by any one of three methods:

1. Pump from after trim to forward trim.

2. Pump from after trim to auxiliary.

3. Pump from auxiliary- to forward trim.

179

d. *All right over-all and all right fore and aft*. The submarine is in neutral

even keel and moving at slow speed, the analysis of fore-and-aft trim can be buoyancy and is in equilibrium fore and aft; hence it requires no trim adjustment.

It may be noted that the above trim conditions might be expressed by different phrasing. For instance, "heavy over-all and heavy aft" might be described as "Heavy over-all and light forward" and "Light over-all and light forward" might be phrased, "Light over-all and heavy aft." However, the standard phrasing is preferred. "Heavy over-all and heavy aft" leaves no room for doubt as to the condition or the remedy. As the object of a standard phraseology is to convey information concisely and efficiently, it should be adhered to at all tines.

18B3. Initial trim. When the submarine leaves the surface it is heavy over-all and heavy forward, the negative tank having been flooded to give negative buoyancy and an initial down-angle. Nearing the specified depth, the negative tank is blown and the planesmen control the angle and depth of the ship. The diving officer now observes the angle of the submarine. If the submarine is heavy or light to such a degree that the planesmen cannot maintain the depth without the assistance of the inclined surfaces of the ship, the inclinometer indicates the corrective action. Should it be necessary to carry an up-angle in order to hold depth, the ship is heavy and ballast is pumped from the auxiliary tank to sea until the planesmen can hold depth on an even keel. In the case of a down-angle, the auxiliary tank is flooded from the sea.

During this first check, which is made immediately on reaching ordered depth,

made. If an excessive angle on the diving planes is necessary to hold a zero bubble, the diving officer orders a readjustment of the ballast in the trim tanks until the planesmen can hold a zero bubble with a minimum use of the planes. As final trim is made at slow speed, any subsequent change in angle requiring excessive use of the planes indicates a further adjustment of fore-and-aft trim; any difficulty in maintaining depth indicates a correction of over-all condition

18B5. Summary. Over-all trim is the process of attaining neutral buoyancy; final trim is the establishment of fore-and-aft equilibrium with neutral buoyancy. The foregoing description divides the procedure into two distinct phases and these are detailed as a sequence of operations. The experienced diving officer will, however, recognize both conditions simultaneously, and combine the operations as dictated by his judgment. Thus he may, in a minimum time, return speed control to the commanding officer by his report, "Final trim," meaning "All right over-all and all right fore and aft."

18B6. Special conditions. When diving, if the submarine, is so badly out of trim that the stern planesman cannot control the angle, the diving officer must first correct the fore-and-aft trim, thus enabling the stern planesman to control the bubble, before attempting any trim analysis. In this extreme condition, it must be remembered that, for a given speed, the forces resulting from the angle of the submarine are greater than those resulting from the angle of the planes. Changes of depth with way on, and carrying an angle, do not necessarily indicate the over-all condition.

the boat is moving at the speed of submergence. As the effects of inclined surfaces are proportional to speed, this first correction may not be enough. The diving officer now orders two-thirds speed and the angle is checked again and corrected as before if necessary. The speed is then reduced to one-third and the check and correction repeated. This trimming operation is repeated until depth can be maintained at slow speed with a zero bubble.

18B4. Final trim . When the ship is on an

Consideration must be given to the forces resulting from the speed and angle of the submarine. Therefore, if the stern planesman cannot control the angle, and the depth is changing in a direction contrary to that desired, the most effective measure that may be employed is that of reducing speed. For example, if the submarine will not go down while carrying all up-angle that cannot be

removed by the stern planesman, it must slow down or stop. If the depth does not increase, then, and only then, may it be concluded that the ship is "light over-all and light forward." Forward trim must be flooded until the stern planesman obtains control of the bubble. When control is obtained, the analysis and correction of trim may be continued.

If the submarine is going down while carrying a down-angle that cannot be removed by the stern planesman, it must slow down or stop. If the depth continues to increase, then it may be concluded that the ship is "heavy over-all and heavy forward." Variable ballast must be pumped from forward trim to sea until the stern planesman can control the angle. It is possible that the condition of trim was "all right over-all and heavy forward." The submarine was in a condition of neutral buoyancy and retained its initial downward motion after the force resulting from speed and angle had been re-

moved. However, this fact did not become known until after the downward motion had been checked.

18B7. Effect of backing while submerged . When the submarine has an excessively large down-angle during submergence, and it is desired to stop its downward motion, it has been stated that part of the force that is driving it down may be removed by stopping the propellers. However, due to the slow rate of deceleration, the downward motion will continue for a time after the propellers are stopped. The quickest way of stopping downward motion is to "stop and back." The reversed propellers tend to force the stern down and the ship squats. This also tends to remove the down-angle. Hence, the procedure for checking downward motion and down-angle when the planes are ineffective is "stop, back, and blow." These measures are resorted to in that order, using one, two, or all, depending upon the emergency.

C. SURFACING

18C1. General principles. Surfacing must be done with caution. The submarine is first brought to periscope depth and a thorough search is made of the surrounding area. When assured that surfacing is safe, the preliminary order, "Stand by to surface," warns the personnel that the signal may be expected.

At the sounding of the signal, three blasts of the diving alarm, or the passing of the word "Surface, Surface, Surface," the various actions necessary are performed.

The bow planes are placed on ten degrees dive and rigged in automatically unless the conning officer gives other instructions. A report, "Bow planes rigged in," is made to the conning officer. Speed is increased to about 6 knots to give maximum lift. Due to the up-angle on the ship, the increased speed makes the inclined surface of the hull effective and the resultant lift raises the ship. The stern planes are used to limit the up-angle to about 5 degrees. The up-angle may be increased by blowing the bow buoyancy tank. Blowing the safety tank increases the

positive buoyancy. However, this is not usually done.

The main ballast tanks are partially blown to surface normally. After surfacing, the high-pressure air is secured and the blow is completed with the low-pressure blowers.

During submergence, certain tanks are vented inboard, and torpedoes may be fired. This causes a consequent rise of pressure within the ship. On occasion this may be a rise of several inches. It has been found by experience that release of the energy in the large volume of air within the pressure hull may result in injury to personnel, damage to hatch gaskets, or the loss of loose gear caught in the rush of escaping air through an open hatch.

Sealing the lower conning tower hatch permits immediate opening of the upper hatch since the relatively small volume of air in the conning tower can be released without danger. This procedure expedites the movement of personnel to the bridge.

When the decks are awash, the conning tower hatch is opened and the commanding officer goes to the bridge. In the meantime,

all stations are alert and prepared to dive at once. The safety of the ship demands that nothing interfere with an emergency dive, should it become necessary. When the commanding officer is satisfied with surface conditions, the announcement, "All clear" will indicate that the submarine

completing the blowing of the main ballast tanks and reducing the pressure within the hull. Usually the pressure is equalized before the lower conning tower hatch is opened. The engine air induction and hull outboard ventilation valves are opened on orders from the bridge. Propulsion is shifted to the main engines.

is to remain on the surface and the remainder of the surfacing routine is carried out.

During this interval, the low-pressure blowers, using air from within the ship, are

The safety tank is flooded, the low-pressure blowers are secured after 15 minutes running, or when the tanks are dry, and normal surface routine is again assumed.

D. PHRASEOLOGY

18D1. Diving and submerged.

a. "Rig ship for dive." The order to carry out the preparations for diving listed in Section 18A4.

b. "Ship rigged for dive," or "Ship rigged for dive except ------------- " A report indicating the accomplishment of the above order.

c. "Clear the bridge." A usual preliminary to the diving signal. All personnel, unless excepted, lay below on the double.

d. "Bleed air." An order to bleed air into the submarine.

e. "Pressure in the boat." The report indicating that the hull is sealed.

f. "Green board." The report meaning that all hull openings are closed, as indicated by the Christmas tree.

g. "Six, (five, etc.) feet." An order to the diving officer, or the bow planesman, giving the desired depth.

h. "Ease the bubble." "Zero bubble." "Five degrees down bubble." Orders to the stern planesman giving the angle desired on the ship.

i. "Shut bow buoyancy vent." "Shut the

m. "Cycle the vents." Given when it is desired to vent safety, bow buoyancy, and the main ballast tanks in succession.

n. "Full rise on the bow planes." "Five degrees dive on the stern planes," Orders to the planesman, giving the angle desired on the diving planes.

o. "Pump from forward trim to after trim." "Flood auxiliary from sea." "Blow from forward trim to sea." "Pump from auxiliary to after trim five hundred pounds." "Secure the pumping." Orders to the trim and air manifold watches, used in shifting variable ballast.

p. "Start the low-pressure blower, blow all main ballast." "Secure the air to number one." "Secure the low-pressure blower." Orders governing the operation of the low-pressure blowers.

q. "Pressure equalized." The report that the air pressure inside the submarine is the same as that of the atmosphere.

r. "Take her down." Increase depth as rapidly as possible. Exact depth will be specified later.

s. "Rig for surface." The order to place the ship in the normal peacetime surface cruising condition.

t. "Open bulkhead flappers, and

main vents." "Open negative flood." Orders governing the operation of the various flood valves and main vents.

j. "Vent safety." Open safety vent, then, after tank has vented, shut the vent

k. "Blow safety." "Blow negative." "Blow all main ballast." Orders to the air manifold watch to blow the designated tank or tanks.

l. "Secure the air." "Secure the air to bow buoyancy." The order to stop blowing all tanks, or the designated tank.

recirculate." Given after the dive has been made and conditions are satisfactory.

u. "Low-pressure blower secured, all main ballast tanks dry, safety and negative flooded, conning tower hatch and main induction open, depth eighteen feet." A typical report by the diving officer, giving the conditions upon completion of surfacing.

19
COMPENSATION

A. GENERAL

19A1. Definition. Compensation is the act or process of counteracting a variable. In connection with a submarine it refers to a redistribution of ballast water to counteract unsymmetrical moment arms and effect balance about the center of buoyancy. It also includes the addition or removal of ballast to offset variations in the over-all weight of the ship.

When submerged, a submarine may be considered to be suspended from a single point called the center of buoyancy. The condition of stable equilibrium requires that the center of gravity be located in a vertical line from the center of buoyancy downward. Following this natural law, a suspended body will always assume this position. In the submarine,

adjustment of ballast to compensate for any change in the over-all weight since the last operation. Further adjustment of the variable ballast is ordered by the diving officer after submergence until perfect trim is obtained and the men at the diving planes steady the ship on an even keel. When the diving officer obtains final trim, the amount of water in each variable ballast tank is recorded in the diving book. On subsequent dives this trim is correct if conditions are the same. However, conditions do not remain static for any extended periods and may change rapidly. On long dives the variable ballast tank readings should be recorded just prior to surfacing. The weights that change on a ship are those of personnel, torpedoes, provisions, fresh water, lubricating oil, fuel oil, and air. Most people, not familiar with

however, this result does not meet the requirements necessary for stable equilibrium on an even keel.

The conditions necessary for equilibrium at even keel follow the laws of the simple balance. The products of the weights on each side of the support times their moment arms, or distance from the support, must be equal. Before a submarine leaves port, it is trimmed for surface operation by the

submarines, do not realize that the compressed air in the air flashes weighs as much as 4 tons on some submarines. These various changes affect the fore-and-aft trim and also the total weight, and must be counteracted by the manipulation of water ballast in the auxiliary and forward and after trim tanks. Compensation therefore is nothing more than balancing; the ship is balanced fore and aft and over-all.

B. COMPENSATION BY MOMENTS

19B1. General method of computing. The proper distribution of ballast is ascertained by consideration of the moment arms of the various weights in question. The center of gravity of the auxiliary tank is assumed to be the center of gravity of the submarine, and the moment arms are calculated from this point to the center of gravity of the variable weights. The table on the next page gives some of these distances for feet-type submarines.

19B2. Examples. Some typical examples will demonstrate the method of compensation by moments.

In the following examples, it is assumed that the ship is in perfect trim and two torpedoes are removed from the racks in the forward torpedo room and placed on the dock. It is desired to compensate by using the forward trim tank and the auxiliary tank.

Two torpedoes weigh 6,354 pounds. The distance of the center of gravity of the forward torpedo racks from the center of the auxiliary tank is 94.5 feet. The distance of the center of gravity of the forward trim tank from the center of auxiliary tank is 113.5 feet. The moment of two torpedoes about the center of gravity is 6,354 X 94.5=

183

	Distance in feet	Location
Torpedoes in forward tubes	118.00	Forward
Forward trim tank	113.50	Forward
WRT tank	98.50	Forward
Torpedoes in forward racks	94.50	Forward
Sanitary tank No. 1	77.50	Forward
Fresh water tanks No. 1 and No. 2	75.00	Forward
Normal fuel oil tank No. 1	68.50	Forward

Battery fresh water tank	67.00 Forward
Normal fuel oil tank No. 2	55.00 Forward
Battery fresh water tank	54.00 Forward
Negative tank	36.50 Forward
Sanitary tank No. 2	24.50 Forward
Stores	23.50 Forward
Fresh water tanks No. 3 and No. 1	19.50 Forward
4-inch magazine	15.50 Forward
Fuel ballast tanks No. 3A and No. 3B	15.00 Forward
Ammunition and refrigeration space	11.50 Forward
Safety tank	7.00 Forward
Auxiliary tank	0.00 Amidships
Battery fresh water tank	2.00 Aft
Battery fresh water tank	14.50 Aft
Reserve lube oil and sanitary tank No. 3	24.00 Aft
Fuel ballast tanks No. 5A and No. 5B	26.00 Aft
Lube oil tank No. 1	36.50 Aft
Main engine sumps No. 1 and No. 2	45.00 Aft
Clean fuel oil tank No. 1	50.00 Aft
Collecting and expansion tanks	62.50 Aft
Lube oil tank No. 2	64.00 Aft
Normal fuel oil tanks No. 6A and No. 6B	69.50 Aft
Main engine sumps No. 3 and No. 4	72.50 Aft
Clean fuel oil tank No. 2	76.50 Aft
Lube oil tank No. 4	80.50 Aft
Normal fuel oil tank No. 7	85.00 Aft
Lube oil tank No. 3	86.50 Aft
Motor and reduction gear sumps	92.50 Aft
Sanitary- tank No. 4	100.00 Aft
Torpedoes in after racks	115.00 Aft
WRT tank	121.00 Aft
Torpedoes in after tubes	139.00 Aft
After trim tank	140.00 Aft

600,453 pounds feet. To obtain the same moment, using the forward trim tank, the torpedo moment is divided by As 9,531 pounds were taken aboard, 1,596 pounds (9,531-7,935) must be pumped from the auxiliary tank to the sea the distance to the forward trim tank: 600,453 / 113.5 = 5,290 pounds.

Therefore 5,290 pounds are flooded into the forward trim tank from the sea, preserving the fore-and-aft trim, but the total weight of the ship would be 1,064 pounds light. This amount, 1,064 pounds, is therefore flooded into the auxiliary tank from sea. and the trim is correct over-all and fore and aft.

The process is the same in more complicated problems as illustrated below A submarine in perfect trim takes aboard 3 torpedoes in the forward racks. Ten thousand gallons of fuel oil are taken into each of the normal fuel oil tanks No. 1 and No. 2. Compensation is desired using the forward trim and auxiliary tanks. Three torpedoes, at 3,177 pounds each, weigh 9,531 pounds; fuel oil weighs approximately 7.13 pounds per gallon; sea water weighs 8.56 pounds per gallon. Using moment arms, it is found that to compensate for the torpedoes, the weight of sea water to be pumped from forward trim tank to sea is:

$$W \times 113.5 = 9,531 \times 94.5$$
$$W = (9,531 \times 94.5)/113.5 = 7,935 \text{ pounds}$$

to maintain this over-all weight. The 10,000 gallons of fuel oil taken into fuel tank No. 1 displace 10,000 gallons of sea water. As the differential in weight is 1.43 pounds per gallon, 14,800 pounds' ballast must be added. The amount flooded into forward trim tank is:

$$W \times 113.5 = 14,300 \times 68.5$$
$$W = (14,300 \times 68.5)/113.5 = 8,630 \text{ pounds}$$

Therefore, 5,670 pounds {14,300-8.630} are flooded into the auxiliary tank. The weight used in the compensation for the No. 2 fuel tank is found to be:

$$W \times 113.5 = 14,300 \times 55$$
$$W = (14,300 \times 55)/113.5 = 6,980 \text{ pounds}$$

Therefore, 6,930 pounds are flooded into the forward trim and 7,370 pounds (14,300 - 6,930) into the auxiliary. The vessel now has the same trim fore and aft and over-all. In actual practice, ballast is pumped or flooded to the nearest point that can be read on the gages.

The ballast disposal may be summarized as shown in the following table:

FORWARD TRIM		AUXILIARY		OVER-ALL WEIGHT	
In	Out	In	Out	In	Out
	7,935		1,596	9,531	
8,630		5,670			14,300
6,930		7,370			14,300
+15,560	-7,935	+13,040	-1,596	+9,531	-28,600
-7,935		-1,596			
					+9,531
+7,625		+11,444			-19,069

			+7,625	
			+11,444	
			+19,069	-19,069

185

The over-all weight may be checked by taking the algebraic sum of all weights added to or taken from the ship, thus

Weights added (in pounds)

9,531 Torpedoes

142,600 Fuel oil

15,560 Ballast in forward trim

13,044 Ballast in auxiliary

+180,731

Weights subtracted (in pounds)

7,935 Out forward trim

1,596 Out auxiliary

171,200 Sea water from fuel tanks

+180,731

180,731 - 180,731 = 0 = net change in weight.

Compensation may also be accomplished by using the after trim tank. The ballast to be pumped from the after trim tank, to equal the 7,625 pounds pumped into the forward trim tank, is:

W X 140= 7,625 X 113.5

$$W = \frac{(7,625 \times 113.5)/140 =}{6,182 \text{ pounds}}$$

Therefore 6,182 pounds out of the after trim tank will maintain the fore-and-aft trim. However the ship is now light over-all and ballast must be added to the auxiliary tank to compensate for the net loss of 19,069 pounds plus the amount pumped from after trim. Total ballast added to the auxiliary would be 19,069 + 6,182 = 25,251 pounds.

C. THE COMPENSATION CURVE

19C1. Description. A more convenient and much shorter method than that given above is by use of a compensation curve. The curve shown in Figure 19-1 is for the submarine herein described. As constructed, the center of the auxiliary tank is the origin, as the center of gravity of this tank is considered coincident with the center of gravity of the ship.

The center ordinate representing the auxiliary tank is laid out in pounds. The axis of abscissas is laid out in thousand-pound feet. The curves for the various tanks are drawn from the data available. A rapid and accurate way to construct the curves is by locating points as far

The forward trim tank curve is found in like manner:

200,000 / 113.5 = 1,762

19C2. Use of curve. In using the curve, the point is taken on the center ordinate that represents the weight to be compensated. It is projected to the litre representing the location of the change. A line is dropped from this point to the line of the forward or after trim tank and from this intersection back to the center ordinate. The point thus found gives the number of pounds to be added or taken from the trim tank to preserve the fore-and-aft trim. If the line is dropped from the intersection on the tank line to the

from the center of origin as possible and drawing the lines to the point of origin.

The curves in Figure 19-1 are located as follows:

The moment arm of the forward torpedo tubes is 118 feet from the center of the auxiliary. The number of pounds required to produce a moment of 200,000 pounds is

200,000 / 118 = 1,695

This point is located on the proper ordinate and the line drawn.

base line, the moment in thousand-pouted feet is read directly.

As the summary of the changes in ballast to effect compensation is the algebraic sum of the weight added to or taken from the vessel, some care must be used to avoid any confusion of signs. In the use of the curves, all tanks and moment arms to the right of the center ordinate are positive and those to the left of the ordinate are negative. Any weight added to or taken from the vessel, which tends to depress the bow or raise the stern, is positive. If the result of

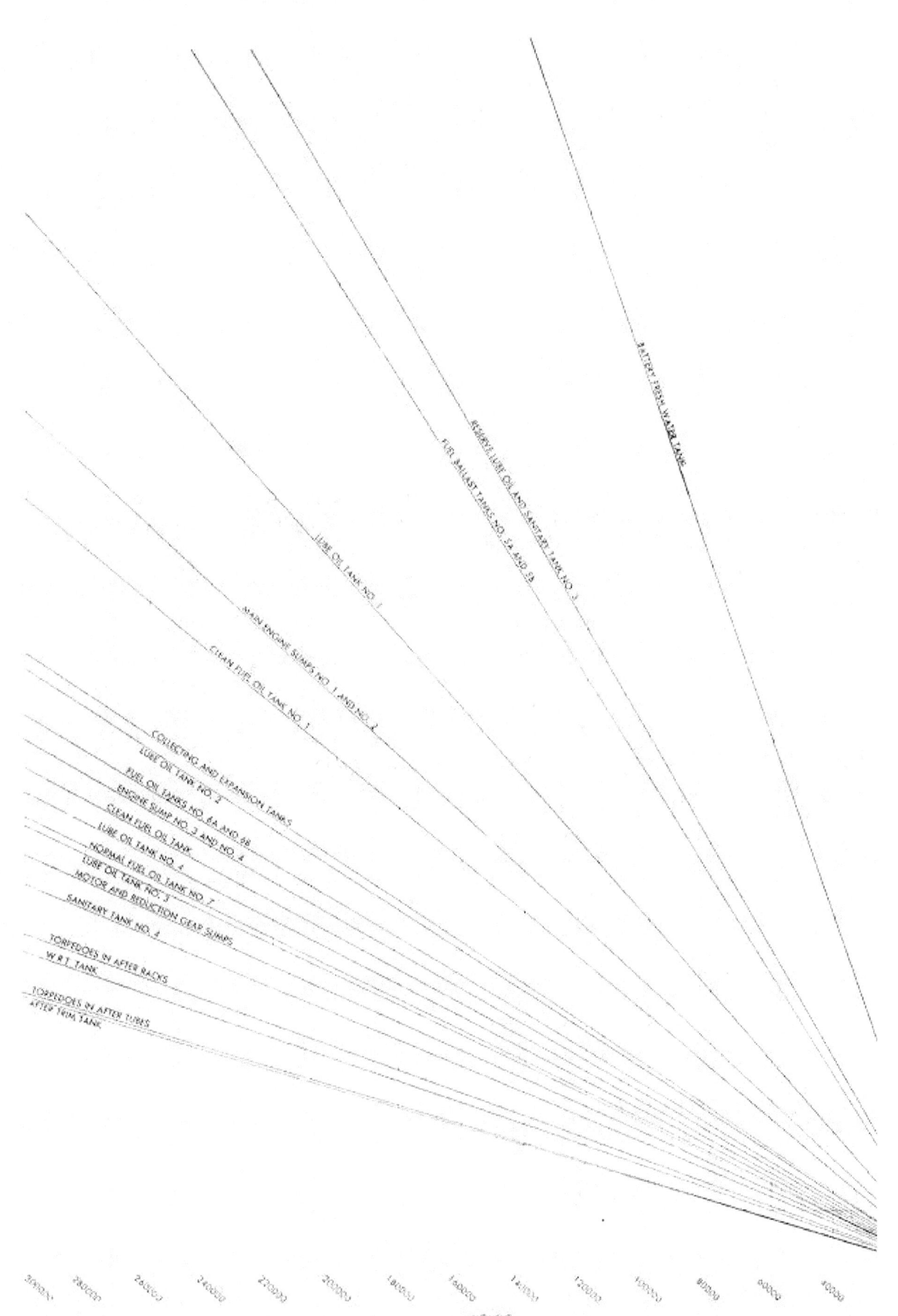

BATTERY FRESH WATER TANK

RESERVE LUBE OIL AND SANITARY TANK NO. 3

FUEL BALLAST TANK NO. 5A AND 5B

LUBE OIL TANK NO. 1

MAIN ENGINE SUMPS NO. 1 AND NO. 2

CLEAN FUEL OIL TANK NO. 1

COLLECTING AND EXPANSION TANKS

LUBE OIL TANK NO. 2

FUEL OIL TANKS NO. 6A AND 6B

ENGINE SUMP NO. 3 AND NO. 4B

CLEAN FUEL OIL TANK

LUBE OIL TANK NO. 4

NORMAL FUEL OIL TANK

LUBE OIL TANK NO. 3

MOTOR AND REDUCTION GEAR SUMPS

SANITARY TANK NO. 4

TORPEDOES IN AFTER RACKS

W.R.T. TANK

TORPEDOES IN AFTER TUBES

AFTER TRIM TANK

300000   280000   260000   240000   220000   200000   180000   160000   140000   120000   100000   80000   60000   40000

MOMENT

the change in weight tends to produce a counterclockwise rotation, the force is negative.

As the charts are laid out with the bow to the right, any change which tends to rotate the boat clockwise about its center of gravity is positive.

The preceding example in compensation is worked out with the curve as follows:

Three torpedoes with a total weight of 9,531 pounds are placed in the forward racks. Ten thousand gallons of fuel oil are poured into the No. 1 fuel tank and 10,000 gallons additional into the No. 2 fuel tank.

As the 9,531-pound weight of the torpedoes extended from the center ordinate does not meet the torpedo rack curve, the weight may be divided until the intersection is on the chart. Dividing by 5 and extending the 1,906 point to the rack curve and from there down to the base line, the moment is read as +180 thousand-pounds feet. Multiplying by 5 gives 900,000 as the moment of the 3 torpedoes.

The fuel oil taken aboard displaces an equal amount of sea water with the result that 14,300 pounds are removed from each of the two fuel tanks. The 14,300 pounds extended runs off the chart, so 1/5, or 2,860, is extended to intersect with the tank curves. Doing this, and extending to the base line,

gives 196.000 for the No. 1 tank and 157,000 for the No. 2 tank. The moments for 14,300 pounds are 980,000 and 785,000.

Summarizing these moments:

+900.000	980,000
	785,000
	1,765,000
-1,765,000	
+900,000	
-865,000	

This shows that the ship is light forward and ballast must be flooded into the forward trim tank to produce an 865,000-pound moment.

As 865 is off the chart it is divided by 5. Projecting upward from 173 on the base line to the forward trim tank curve and then left to the center ordinate, the amount is read as 1,520. Multiplying by 5 gives 7,600 pounds as the amount to be flooded into the forward trim tank.

As 9,531 pounds were added to the vessel and 28,600 pounds were lost as a result of the displacement of sea water by the fuel oil, the net loss in over-all weight is 28,600 - 9,531 = 19,069 pounds. Since 7,600 pounds were added to the forward trim tank, 11,469 pounds must be flooded into the auxiliary tank to maintain the original over-all trim.

D. COMPENSATION BY PERCENTAGE

19D1. Description. Another type of compensation curve may be constructed using percentages (Figure 19-2). The center of the auxiliary tank is the origin. This ordinate is laid off in percent, 0 to 100. The base, or line of abscissas, is laid off in feet and the relative locations of the centers of gravity of the different tanks are indicated. Lines are drawn from the 100-percent point in the center ordinate to the locations of the forward and after trim tanks.

19D2. Use of curve. This curve is used by projecting the point, indicating the tank in which the change in weight is made, up to an intersection with the percentage curve. This intersection gives the percentage of the weight which must be pumped out of or flooded into the variable tanks to correct the fore-and-aft trim and over-all weight. This intersection is projected to the center ordinate. This point gives the percentage of the weight in question which must be pumped out of or flooded into the auxiliary tank. When projected to the corresponding trim tank scale it gives the percentage of weight to be flooded into or pumped out of the trim tank.

The previous example may be worked out with this curve as follows:

Weight in forward torpedo racks is 9.531 pounds

Weight out (net) No. 1 fuel tank is 11,300 pounds

188

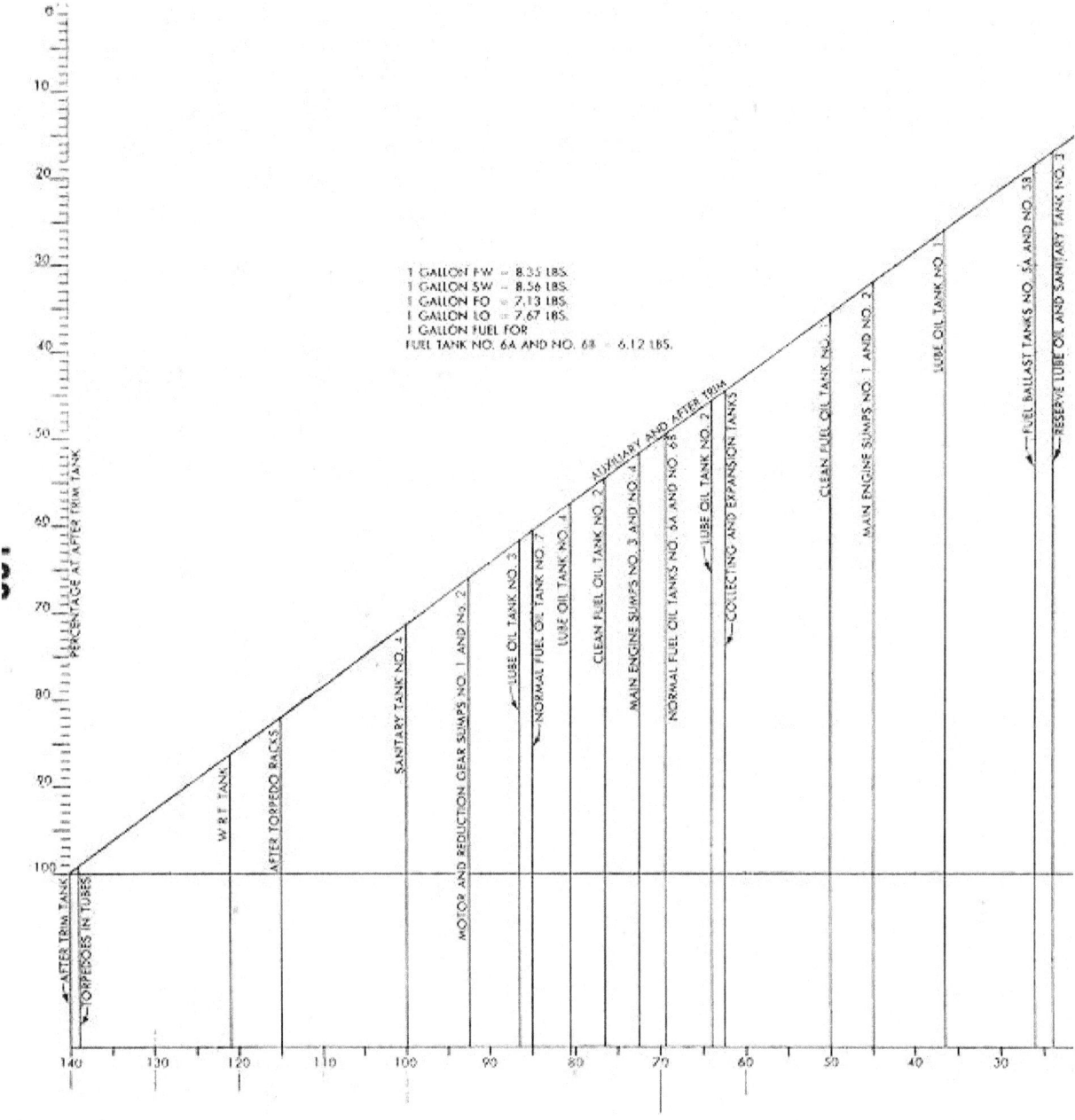

Figure 19-2. Compensation by percentage

189

Weight out (net) No. 2 fuel tank is 14,300 pounds

1,573 pounds out of auxiliary.

The 14,300 pounds out of No. 1 fuel

The forward torpedo rack intersection with the forward trim tank line shows that 83.5 percent of the 9,531 pounds should come out of the trim tank and the remainder, 16.5 percent, out of the auxiliary.

7,958 pounds out of trim tank.

tank reads 60.5 percent or 8,650 in the trim tank, the remaining 39.5 percent of 5,650 in the auxiliary tank.

The 14,300 pounds out of No. 2 fuel tank reads 48.5 percent or 6,935 pounds in the trim tank, the remaining 51.5 percent or 7,365 pounds in the auxiliary.

SUMMARY											
TANK OR SPACE	PERCENT			WEIGHT CHANGE		FORWARD TRIM		AUXILIARY		AFTER TRIM	
	F.T.	AUX.	A.T.	Lighter	Heavier	In	Out	In	Out	In	Out
Fwd. Racks	83 1/2	16 1/2			9,531		7,958		1,573		
No. 1 F.O.	60 1/2	39 1/2		14,300		8,600		5,650			
No. 2 F.O.	48 1/2	51 1/2		14,300		6,935		7,365			
				28,600	9,531	15,585	7,958	13,015			
				9,531		7,958		1,573			
				19,069		7,627		11,422			
						11,442					
						19,069					

Pumping 7,627 pounds into the forward trim from the sea and flooding 11,442 pounds into the auxiliary, trims the ship fore and aft and over-all. Pumping and flooding to the nearest 50 pounds, as is the practice, the figures would be 11,500 pounds into the auxiliary and 7,600 pounds into the forward trim. The ship is 27 pounds lighter forward and 31 pounds heavy over-all.

If the location of the weight change is in the after part of the ship, compensation is effected by using the after trim tank and the auxiliary tank.

The computations necessary for any problem in compensation may be quickly and easily made by the use of either of the curves. It is of the utmost importance, however, that extreme care be used to include all the weight changes and to get them in the proper columns. Small mistakes in the case of some tanks may make a large error in the final trim. For example, a 100-pound mistake in the weight change in the forward trim tank has the same effect on the fore-and-aft trim as a 1,040-pound change in the ammunition and refrigeration space.

PATROL ROUTINE

A. INTRODUCTION

20A1. Foreword. The following patrol instructions are a compilation of instructions used by various submarines. Naturally, there may be some minor differences between these instructions and those used by a particularsubmarine.

B. DUTIES OF WATCH STANDERS

20B1. Officer of the deck. On the surface, the officer of the deck stands his watch in the forward bridge structure. Although he is expected to remain intensely alert and observant, he is not a lookout and must not become engrossed in a detail of his watch or a lookout sector to the exclusion of his comprehensive duties as supervisor of the watch. His responsibility when the ship is submerged is no less than when on the surface. and a similar degree of alertness is required in carrying out the routine and direction of the watch. The duties of the officer of the deck as outlined in *Navy Regulations*, are supplemented as follows:

Keep the number of persons on the bridge to a minimum, requiring permission to come on the bridge in each case.

Allow only one relief of any watch on the bridge at a time, with the exception that besides one lookout relieving, the quarter master or junior officer of the deck may relieve.

Insure yourself that the lookout's vision has become dark-adapted before allowing him to relieve the watch. A

Keep them as dry as possible and out of high wind. A comfortable lookout is much more efficient than an uncomfortable one.

The following rules apply generally, but in no way restrict the officer of the deck from acting as his judgment dictates:

Dive for all aircraft contacts, except as specifically directed by previous instructions of the commanding officer.

Turn toward a periscope forward of the beam and go to full speed. Turn away from a periscope abaft the beam and go to full speed.

Turn away from all fishing vessels or small craft unless ordered by the commanding officer to attack.

Turn away from unidentifiable objects.

Turn toward a target, but dive in sufficient time to insure that your ship is not sighted prior to firing torpedoes.

Present the smallest target possible by turning toward or away from any type of contact as the situation dictates.

reasonable test is the lookout's ability at a distance of about 5 feet to note how many fingers you have extended.

Keep the lookouts alert and insure that they are properly covering their sectors. Insist on standard phraseology in all reports, with prompt acknowledgments. Maintain the passageway to the hatch clear at all times.

Insure that the quartermaster orders rainclothes for the watch in sufficient time to permit one person at a time to don them prior to arrival of a squall. Place the lookouts where they will be of most advantage.

In friendly waters, or when contact with own forces is probable, have daily recognition signals written in chalk on wind screen.

Know and insure that all bridge personnel on watch know the current signals in effect, and that recognition gear is in complete readiness as follows:

1. Searchlight tested if rigged.
2. Blinker tube readily accessible.
3. Flares and rockets changed at proper times.

Carry out the following details or routine:

Approximately 15 minutes before diving secure the 20-mm guns and ammunition.

Carry out the diving procedure.

After daylight and torpedo routining,

the depth at which to run during and between looks will depend on the state of the sea and the proximity of enemy air bases, as directed by the commanding officer.

When landmarks are available, keep the ship's position cut in, using periscope exposures of short duration only.

In making periscope observations, first sweep the horizon and sky in low power for aircraft or close surface craft. Follow this, if all clear, by a slow deliberate search in high power, not exposing the periscope for more than 15 seconds. Develop the habit of obtaining a complete picture of the weather during your observation. Have the quartermaster note the keel depth

Obtain the commanding officer's permission prior to the following:

1. Permitting anyone on deck.
2. Putting any piece of machinery or armament out of commission.
3. Pulling a torpedo from any tube.
4. Anything that may reduce the fighting ability of the ship or her ability to dive.

For every sighting while on the surface that might develop into an attack, *sound the general alarm* and let the rest of the ship know something is happening.

Whenever the need arises to make ready torpedoes, order " *Make ready tubes forward (or aft).*"

Keep the junior officer of the deck

during the observation and also tell you when 15 seconds of periscope exposure has occurred.

When the target is sighted, sound the general alarm, or pass the word over the telephone if the target is likely to hear the alarm, and commence the approach immediately. Minutes or seconds may be just as valuable then as later in the attack, and certainly are if the target is presenting a large angle on the bow.

Habitually require a smart trim from your diving officer for the depth at which you are running.

Notify the commanding officer when darkness is almost ready to set in and obtain the time of surfacing from him. Commence the surfacing procedure.

Ten minutes prior to surfacing, the commanding officer comes to the conning tower and relieves the watch. The junior officer of the deck of the previous watch relieves the diving officer a half hour prior to surfacing. Normally, just prior to surfacing, the ship is brought to 50 feet for SD sweep, then to 40 feet for SJ sweep (sound sweeping all the time).

If "All clear on radar" is indicated, pass. the work over the loudspeaker, " *Stand by to surface .... engine combination.*" The commanding officer will then direct the surface alarm be sounded.

Carry out surfacing procedure on the surface, alarm.

informed of changes of speed, course, or any other pertinent information so that he will be able to assume the deck at any time.

20B2. Junior officer of the deck. The junior officer of the deck stands his watch as directed by the commanding officer. He may act in any one of three capacities as follows:

As junior officer of the deck, stationed forward or aft the cigarette deck. (If aft, quartermaster should go forward.)

As junior officer of the deck, stationed in the conning tower.

The junior officer of the deck watch normally is stood only while on the surface. Until all officers are fully qualified as diving officer, the junior officer of the deck normally acts as diving officer submerged. However, if qualified by the commanding officer, he may interchange with the OOD for periscope watches if practicable.

Duties of the junior officer of the deck are as follows:

Supervise the lookouts to insure that they are covering their sectors properly.

Observe to insure that any enemy that might possibly get by the lookouts does not approach, unobserved, to close range. (If forward, observe from broad on opposite bow to stern.)

Note the condition of the 20-mm guns and be prepared to man or direct the fire as directed by the OOD.

Report immediately to the officer of the deck own engine smoke, sparks, or any unusual condition.

Proceed to diving station or below on orders of the officer of the deck or on " *Clear the bridge.*" Carry out the diving procedure. Conduct the periscope watch as directed. Man the TDC and otherwise assist the OOD in conducting initial stages of any approach as directed.

The junior officer of the deck should be fully aware of the condition of the boat and prepared to assume the capacity of the officer of the deck at any time.

20B3. Conning tower talker watch. The conning tower talker watch is normally stood at all times when at sea. Conning tower talker watches are stood as indicated on the Watch, Quarter, and Station bill.

The duties of this watch may be outlined as follows:

Instruct new steersmen in their duties.

Make all necessary entries in the Quartermaster Notebook.

Act as voice link between sound, radar, control, maneuvering, and bridge watches.

Supervise zig plan if in use.

Take and record TBT and heading bearings on bridge buzzer marks.

Act as telephone watch.

Keep conning tower clear of loose gear.

The duties of the steersman are as follows:

Maintain the course.

Operate the maneuvering room annunciators as ordered by the bridge.

Know the duties of the conning tower watch.

Assist the conning tower watch as necessary and be able to take over at any time.

Know exactly where all conning tower alarms are and operate them only when ordered to do so by the bridge.

20B5. Quartermaster of the watch. The quartermaster of the watch normally is stationed on the bridge, aft when cruising, and may exchange with the OOD if ordered. He is an additional all-around lookout and does not restrict his search to any one sector unless so ordered by the officer of the deck.

The quartermaster is responsible under the direction of the OOD for the following routine duties:

Break out binoculars, dark glasses, proper flares, and blinker tube prior to surfacing; also issue lens paper to lookouts.

Obtain warmer clothing or rainclothing for lookouts.

Change flares at the proper time.

Check TBT's upon surfacing each night.

Wipe the periscope windows on surfacing and 15 minutes before routine

Under supervision of the chief petty officer of the watch, permit but one relief for any watch to proceed to the bridge.

At night, check and carry out the night orders.

Upon surfacing, check TBT's with the quartermaster as directed.

Assist the quartermaster in maintaining deck log columns.

Maintain quiet in the conning tower and keep an alert watch. Report to the OOD when the conning tower watch is properly relieved.

Know exactly where all the conning tower alarms are and operate them only when ordered to do so by the bridge.

20B4. The steersman. The steersman is normally stationed in the conning tower, unless otherwise ordered by the officer of the deck.

dives.

Operate the periscope, keep the periscope officer informed of depth, and read and record bearings when submerged.

Keep the conning tower clean, and all gear properly stowed when submerged.

Check columns of deck log after being relieved to make sure that the proper entries have been made. Do this in the control room.

Once every hour on surface, check the lookouts' glasses for cleanliness and proper setting.

20B6. Chief petty officer of the watch. The CPO of the watch remains in the control room. He is charged with running the below deck routine, supervising the control room watch when on the surface, and with carrying out the details of the Watch Bill. He initiates the diving procedure on the diving alarm, until relieved of the dive by the

diving officer. In carrying out his duties, he must pay particular attention to the following:

Call the oncoming watch in sufficient time for them to relieve 15 minutes before the hour, in accordance with naval custom.

At night, insure that each oncoming lookout is fitted with and wears dark adaptation goggles continuously for at least 20 minutes before being allowed to proceed to the conning tower.

Insure that only one relief proceeds to

2. Pump all the bilges to the sea.
3. Blow all the sanitary tanks.
4. Collect all trash and garbage and when properly sacked, report to the bridge, "All trash and garbage assembled," and dump it when directed.

Carry out evening compensation as directed by the diving officer.

When orders to the steersman for changes of course or speed come from the bridge over the system, observe the motor order telegraph repeater or rudder angle indicator in the control room to check that the order is being carried out

the conning tower at a time.

Promptly acknowledge any orders or word passed from the bridge or conning tower.

Periodically check the compensation by liquidometer gages.

Insure that the proper watch is maintained on the control room (SD) radar when ordered manned, and that any contact, however doubtful, is reported instantly to the, OOD.

Maintain quiet and allow no loitering in the control room.

Half an hour prior to surfacing, rig the hatch skirt. Turn out the white lights; turn on the red as designated for the control room.

Execute the 2200 lights out in the crew's Mess.

At the end of each watch, and approximately 1 hour before diving, pump the bilges and blow the sanitary tanks to the sea.

Keep the manometer needles matched on surface.

Keep submerged identification signal available as directed by OOD.

Maintain the air banks at proper pressure.

See that all the topside reliefs are properly clothed.

Keep the control room clear of all loose gear.

properly.

Keep the compass check book, making entries as required by the navigator. Instructions will be posted in the front of the book. While submerged, whenever word is passed from the conning tower, "Man battle stations," dispatch one man forward and one man aft with the order, "Pass word quietly, wake all hands. Battle stations submerged."

20B7. Lookouts. Normally there are three lookouts assigned overlapping sectors as follows:

Starboard lookout 350 degrees - 130 degrees (relative).

After lookout 120 degrees - 240 degress (relative).

Port lookout 230 degrees - 010 degrees (relative). In the event that four lookouts are used, sectors are assigned as follows:

Starboard forward lookout 350 degrees - 100 degrees (relative).

Starboard after lookout 080 degrees - 190 degrees (relative)

Port after lookout 170 degrees - 280 degrees (relative).

Port forward lookout 010 degrees - 260 degrees (relative).

During daylight, each lookout searches his sector in the following sequence using a sun filter only when searching into the sun:

a. Search the water to the horizon for one-half of his assigned sector.

About 1 hour before surfacing when directed by the OOD, pump down the *pressure in the boat* to one-tenth.

After surfacing, carry out the evening routine which consists of the following:

1. After the blowers are secured, start. air change.

b. Lower the binoculars for approximately 10 seconds to survey entire sector, water, and sky, with naked eye. Continue search of water to horizon over the remainder of the sector. Search the horizon and lower sky for one-half of the assigned sector. Lower binoculars for approximately 10 seconds to

survey the entire sector, water and sky with the naked eye. Continue search of the horizon and lower sky over the remainder of the sector. Repeat 10-second sweep of the entire sector with naked eye. Search the upper sky, above the belt observed when searching the horizon and lower sky, for one-half of assigned sector. Lower binoculars for approximately 10 seconds to survey the entire sector, water and sky, with naked eye. Continue search of upper sky for remainder of sector. Repeat 10-second sweep of entire sector with naked eye. Recommence, starting with (a) above.

During darkness, the search will be as follows:

a. Moonlight nights when enemy air search is possible: After each complete sweep of sector, search sky sector with naked eye.

b. Dark nights: Eliminate sky search.

This method of search has the following advantages:

It provides a systematic coverage of the entire area.

It gives maximum insurance against any plane, which was outside the field

acknowledged and keep on reporting until you get an acknowledgment.

Do not take eyes or binoculars off the object you have sighted.

Report everything.

Upon assuming your post after surfacing, make a complete search of your sector. Report in a loud clear voice, "......... sector all clear, sir."

At night, don't attempt night duties until dark-adapted; avoid short cuts. Practice use of the corners of the eyes, remembering that objects are better seen in dim light if not located in the center of vision. Move the eyes frequently, remembering that night vision is most sensitive immediately after the line of sight has been shifted. When relieving, make certain that no other bridge watch is being. relieved, then request, "Permission to come on the bridge to relieve lookout." Relieve with a minimum of noise and confusion. Get dressed below for the existing weather conditions.

20B8. Sound watch. a. *General.* The sound watch normally is stood whenever speed conditions permit. The operator must, without seeking confirmation or

of the binoculars, closing unobserved to close range.

Provides best assurance that a periscope lowered during the binocular search, will be sighted if dangerously close during the naked eye sweeps.

General instructions to lookouts:

Save your eyes. All lookouts should rest their eyes before coming on watch. They should try to take care of all calls of nature before going on watch.

If a lookout does not feel physically up to standing an all-out lookout watch, he should report this to the OOD.

Make all reports of sightings immediately. It is better to be wrong 100 tunes than miss one ship.

Use relative bearings in all reports. Then, followed by your best estimate of the range, add more information as it becomes available, stating identity of ship and so forth.

Call out your reports so that all can hear. Make certain your report is

help from anyone, report immediately to the officer of the deck any echo ranging, propellers, or unusual sounds.

b. *Instructions for standing sound watches.* The safety of the ship and its personnel is directly dependent upon the manner in which this watch is stood. This responsibility is greatly increased at night and a resultant increase in attentiveness is imperative.

This watch must be stood in regulation manner. *Submerged*, each man upon being. relieved reports to the officer of the deck, *"Sound watch relieved by ...., Sound conditions are (good, fair, poor)." On Surface*, report to the conning tower talker that you have been properly relieved.

Soundheads are to be used in accordance with communication officer orders posted at the sound gear.

If for any reason, you have difficulty interpreting what you hear, or the equipment

does not appear to be operating correctly, inform the officer of the deck at once, and call for one of the battle station soundmen at the same time without any further orders.

c. *Additional information regarding night sound, watches.* Soundheads should not be left lowered above 10 knots.

When two soundmen are on watch at

Each soundhead should be rotated 360 degrees on alternate sweeps.

If screws are heard, they are to be reported immediately, stating " *Screws at .... relative, (high or low) speed."* Then obtain the closest true bearing and report, " *True bearing ...."* Thereafter, report any change in speed of screws, and if you can no longer hear them. Changes of bearing when own ship is on a steady course are very important. *Keep the*

the same time, both soundheads are lowered. The starboard (JK) operator covers the sector from zero to 180, and the port (CQ) operator covers the sector from 180 to 360.

information coming.

C. DIVING AND SURFACING PROCEDURES

20C1. Diving procedure.

A. Officer of the deck.
   1. Pass the word, " *Clear the bridge.*"
   2. Check all hands below.
   3. Sound two blasts on the bridge diving alarm as the last lookout passes the OOD.
   4. When below, check report Pressure in the boat and order the depth desired.
   5. Commence attack, evasive tactics, or rig for depth charge as conditions warrant.

B. Quartermaster.
   1. Be last down hatch and shut hatch.

C. Junior officer of the deck.
   1. Be among the first down the hatch if on the bridge.
   2. Proceed immediately to diving station.
   3. Assume control of dive, carrying out procedure listed under control room.

D. Lookouts.
   1. Clear the bridge on the double man diving stations in accordance with the Watch. Quarter, and Station Bill.

E. Conning tower talker watch.
   1. Sound the general alarm *if ordered by the OOD.*

   2. Ring up full speed on the annunciators.

G. Maneuvering room.
   1. Answer bells as ordered.
   2. Shut the maneuvering room induction hull valve.

H. Engine room. Procedure in the engine room is carried out in this order
   1. Stop the engines.
   2. Shut the outboard exhaust valves.
   3. Shut the engine and supply ventilation induction hull valves.
   4. Shut the bulkhead flappers.
   5. Shut the inboard exhaust valves.
   6. If everything is in order in the engine rooms-oilers to control room-for submerged stations.

I. Control room.
   1. Open the vents.
   2. Rig out the bow planes and put them on 22-degree dive.
   3. When all outboard exhaust valves are shut or when passing 23 feet shut the engine air induction and the ship's supply outboard valve.
   4. Bleed air in the vessel and secure air on orders of the diving officer only in case the torpedo rooms do not bleed in air.
   5. Note and report pressure in boat.
   6. Put the stern planes on dive to take the ship to the ordered depth with a 4- to 6-degree angle.

2. Lower the periscope if it is up.

F. Helmsman.
    1. Put the rudder on as ordered; otherwise amidship.

7. Shut the bow buoyancy at 30 feet and open the safety for 5 seconds; then shut it. Shut all vents at 50 feet.
    8. Blow the negative to 1,500 pounds in two steps.
    9. Reduce the speed and adjust the trim as necessary.
    10. Shut the negative flood and vent tank inboard.
    11. Make certain that the periscopes are cut in.
    12. When convenient, open all bulkhead flappers and resume normal hull ventilation.

J. Crew's mess.
    1. Report the engine air induction and hull ventilation valves shut by hand signal to control room.
    2. Lock the above valves shut as soon as possible and report to control.

K. All stations.
    1. Shut the bulkhead flappers.

L. Forward torpedo room and after torpedo room.
    1. Bleed air into the vessel and secure on word from the control room.

M. Radio room.
    1. Disconnect the antenna lead and shut the trunk flapper.

The conning officer retains speed control at all times, However, this will not interfere with or necessitate any hesitation by the diving officer to

7. On surfacing, be prepared for immediate diving should circumstances warrant such action.
    8. Upon proceeding to the bridge, the OOD takes the starboard side and searches, the JOOD takes the port side, and the QM the after half of ship. When all three have reported, "All clear," call to conning tower, "Routine."
    9. When the word, "Routine" is passed, the officer in the conning tower sees that the following procedure is carried out:
    a) Main induction opened.
    b) Lookouts to the bridge.
    c) Engines are automatically started on opening of induction.
    d) Turbo blow for 6 minutes.
    c) Gunner's mate to bridge to rig 20 mm.
    f) Rig in soundhead if going two-thirds speed.
    g) After steps a) to e) have been carried out, announce to control, "Carry out all routine below."

b. *Helmsman.*
    1. When the surfacing alarm is sounded, ring up two-thirds speed unless otherwise ordered.

c. *Quartermaster.*
    1. Stand by the hatch.
    2. Sound the surfacing alarm *on orders of the C.O. only.*
    3. Open the hatch on orders from the C.O.

request speed changes to facilitate depth control.

20C2. Surfacing procedure.

a. *Duties of the officer of the deck.*

1. Prior to surfacing, carry out surfacing routines as outlined in the front of the QM notebook.

2. Have the sound watch make an exacting search.

3. Keep the periscope watch as directed.

4. Give control any changes in ordinary surface procedure.

5. Pass the word, "Stand by to surface .... engine combination."

6. Follow the commanding officer and the quartermaster to the bridge. (The JOOD follows the OOD to the bridge.)

d. *Assistant navigator (quartermaster).*

1. Keep the C.O. informed of keel depth and pressure in boat.

e. *Control.*

1. Start the hydraulic plant at 50 feet.

2. Rig in the bow planes when the word is passed to stand by to surface .... engine combination.

3. Blow the bow buoyancy and main ballast, except the safety, on the third blast of the surface alarm. Surface with 3-4 degree rise angle and secure the air when at 30 feet.

4. When the conning tower hatch shows a red light, and the hatch is heard

to open, vent and flood the negative. Vent and shut the safety.

5. Open the main induction on orders from the conning tower.

6. Put the low-pressure blowers on tank as directed.

7. CPO of the watch: Carry out routine as directed. "Carry out all routine below" means:

a) Have the battery charge started.

b) Blow all the sanitary tanks.

c) Pump all the bilges in succession from aft to forward.

d) Assemble trash and garbage in control room.

c) Speed up the exhaust and supply blowers.

f) Start the air change after stopping the turbo blow.

g) Swab down and clean all compartments.

h) Report to the bridge that "All routine is being carried out below."

20C3. Check-off list prior to surfacing.

a. Procedure 30 minutes prior to surfacing

1. Pump down the pressure in the boat to 0.1 inch upon orders of the C.O.

2. Start the hydrogen detectors and line up the battery ventilation for charge. Take individual cell reading.

3. Call the lookouts and have them dress appropriately for the weather.

4. Quartermaster: Clean all binoculars and get dressed to go to the bridge.

5. Bring .45-caliber pistol and submachine gun to the conning tower with two drums of ammunition.

6. Get readings on all sanitary tanks, bilges, and fresh water tanks and record them for final trim estimate.

7. Rig the curtain in the control room. Darken the control room and conning tower.

8. Be prepared to dive again immediately.

f. *Crew's mess.*

1. Unlock and put the hull ventilation supply and engine air induction valves on *power*, on "Stand by to surface=." Report to control when this is accomplished.

g. *Maneuvering room.*

1. Answer bells as ordered. After the surface alarm is sounded, shift to surface rpm.

2. Ring up "Start" on the engines desired.

3. Shift to engines for propulsion when the engine room is ready and start the battery charge as soon as possible.

4. Report to the conning tower as soon as the battery charge is started.

h. *Engine room.*

1. When "Start" is received from the maneuvering room *and* when the outboard engine air induction valve is opened, start the engines as ordered, Carry out the normal routine.

b. *Procedure 15 minutes prior to surfacing.*

1. Notify the maneuvering room and engine rooms to stand by engine combination in accordance with night orders.

2. Check to ascertain that all lookouts are in the control room dressed to go on watch with dark glasses.

3. Navigator: Relieve the OOD to get dressed for surfacing.

4. Diving officer on proceeding to watch: Relieve the diving officer to get dressed for surfacing and the JOOD watch.

5. Trim manifold man: Relieve the bow planesman to get dressed for surfacing.

6. Start up the SJ radar and have the radio technician man the SJ radar in the conning tower.

c. *Procedure upon surfacing.*

1. Immediately upon surfacing, stern planesman and trim manifold man go to the engine rooms.

D. APPROACH OFFICER

20D1. Approach officer's check-off list.

a. Make torpedoes ready early, Set the depth and speed on the torpedo.

b. Give the identification officer all necessary information early.

c. Slow sufficiently for safe periscope observation.

observations for air patrol and other surface patrol craft.

e. Make observations frequently.

f. As range decreases, run at deeper depth.

g. Keep the ship informed of progress of approach as consistent with circumstances.

d. Make sufficient all-around   h. Rig the ship for depth charge attack.

E. TORPEDO ROOMS

20E1. Instructions for torpedo rooms.

a. At least one qualified torpedoman must he on watch in each torpedo room at all times. The phones should be manned continuously.

b. Gyro spindles should be engaged at all times, and in the absence of other instructions, torpedoes set on 0 degrees forward and 180 degrees aft.

c. Unless otherwise instructed by proper authority, the condition of the gyro regulators should be:
    1. Clutch in *automatic* and switch *on.*

d. All torpedoes are to be set for high speed and 10-foot depth until ordered changed from the conning tower. Depth setting spindles should be left disengaged.

e. When the order, "Make ready the tubes forward (aft)" is received, all tubes in the nest must be made ready as rapidly as possible and, as far as possible, in numerical order.

f. Unless word is received to the contrary, the firing order is always in numerical order.

g. Tubes should always be flooded from WRT. In case of an emergency, when submerged, that prevents tubes from being flooded from WRT, request permission from the diving officer to flood the tubes from the sea. If on the surface, notify the diving officer, and flood from the sea.

h. The senior torpedoman in each room is responsible for these orders being properly carried out.

F. STANDARD PHRASEOLOGY

20F1. Examples of standard phraseology.
The following is the standard phraseology to be used on phones and when passing the word:

a. Torpedoes.
    1. Make ready the bow (stern) tubes.
    2. Set torpedo depth 10 feet.
    3. What is the torpedo run?
    4. What is the gyro angle?
    5. Do the bow (stern) tubes bear?
    6. Check fire (or cease fire).
    7. When gyro angle is zero, commence firing.
    8. Use one (six) torpedoes.
    11. Stand by one, and so on. Fire one, and so on.
    12. What is the torpedo track angle?

b. *Own ship-fire Control.*
    1. What is the pit log speed?
    2. Pit log speed is ....
    3. What is our head(ing) (course)?
    4. Our head(ing) (course) is ....
    5. Left (right) 5 degree rudder.
    6. Rudder is 5 degree (10 degree) (full), right (left).
    7. Depth 60 (etc.) feet.
    8. Emergency, 200 (150) feet.
    9. Take her down.
    10. Rig ship for depth charge attack.

9. Use angle, speed, bearing (periscope) spread.

10, Set gyro regulators in hand (automatic).

199

12. What is the log distance?

c. *Target.*

1. (Generated) angle on the bow is .... starboard (port).

2. Estimated (generated) (radar) (sound) range, five oh double oh.

3. Estimated target speed .... knots (low 2-10) (moderate 10-15) (high 15-25)

4. Target is maru (destroyer) (man-o-war) etc.

5. Target (relative) (Sound) (Radar) (True) (generated) bearing is ....

6. Estimated (generated) target course is ....

7. Stand by for a set-up.

8. Target estimated length is ....

9. Target track angle is ....

10. Set-up is good. (T.D.C. set-up.)

d. *Own ship: administration.*

1. Put low-pressure blowers on all main ballast tanks for .... minutes.

2. What time is it?

3. Rig ship for depth charge attack.

4. Forward room rigged for depth charge attack.

5. Open all bulkhead flappers, resume normal hull ventilation.

6. Get ready on the engines.

7. Shift to the battery.

8. Man the sound gear. (Station the sound watch.)

11. Stand by to dive (clear the bridge).

9. Man the radio room (conning tower) radar.

10. House the underwater log.

11. Rig out the underwater log.

12. Lower and lock the port (starboard) soundhead.

13. Raise and lock the port (starboard) soundhead.

14. Open the main induction, start the blowers.

15. Lookouts to the bridge.

e. *Orders to the men on the lines.*

Slack off: Pay out the line, allowing it to form an easy bight.

Take a strain: Put the line under tension.

Take in the slack: Heave in on the line, but do not take a strain.

Ease it, Ease away, or Ease off: Pay out enough to remove most of the strain.

Check Number ....: Hold the line, but render it enough when necessary so that it will not part.

Hold Number ....: Take sufficient turns that the line will not give.

Double up and secure: Run any additional lines and double all of them as necessary.

Stand by your lines: Man the lines, ready to cast off.

Cast off Number ....: Let go Number .... line.

200

SUBMARINE TRAINING DEVICES

A. GENERAL

21A1. Introductory. The modern feet-type submarine is an exceedingly complex mechanism. On the surface, in normal operation, it presents all the problems of ship handling and navigation common to surface vessels. Its problems do not end there, however, for when it submerges it becomes, in effect, an entirely different vessel with new characteristics and new problems.

Submerged, the control becomes more complicated; the ship must be navigated in a three-dimensional medium and many conditions affecting its operation are much more critical than when it is on the surface. Added to these complexities are the more limited facilities for observation and the necessity of relying, to a great extent, on dead reckoning.

As the submarine is an offensive weapon, its chief value against the enemy is its ability to approach undetected and to maneuver to a firing position despite the target's efforts to avoid contact. The approach and attack phase of submarine warfare is a science in itself, requiring a practiced eye, an analytical mind, and the ability to make swift and accurate decisions. Though the possession of these attributes is a paramount requirement for submarine officers, to be of value in submarine attack they must be supplemented with long experience and thorough training.

In the early days of the submarine this experience and training were acquired through actual service under a competent commanding officer. With the growth of the fleet and the phenomenal development of the submarine and its equipment, it became evident that qualified personnel must be secured in ever increasing numbers and trained more rapidly. To this end, numerous training devices, duplicating the more important features of a submarine, have been developed and may be used to simulate actual situations encountered or patrols.

Three departments of the submarine have been the object of particular attention, the conning tower, the control room, and the torpedo room, resulting in the production of three devices known as the *attack teacher*, the *diving trainer* and the *torpedo tube trainer*. These devices are used to train fire-control parties, diving officers, and control room personnel, and to instruct in the care and manipulation of torpedoes and torpedo tubes and in the firing of torpedoes.

B. THE ATTACK TEACHER

21B1. Description . The attack teacher is a device by which typical approach and attack problems may be duplicated in all their phases. The fire-control party in training is assembled in a the conduct of an actual operation against an enemy.

Early attack teachers bore little resemblance to the devices of the present mock-up conning tower. Miniature models of enemy vessels are maneuvered in the field of a specially designed periscope and the fire-control party simulates day. Like the submarines, they have developed rapidly, and they now afford a reliable presentation of battle problems and facilities for their solutions.

C. THE DIVING TRAINER

21C1. Description. The diving trainer (Figure 21-1) consists of a duplicate of the port side of the control room with all the apparatus and equipment usually installed in that section of the submarine. The control zoom section is mounted so that it may be tilted to assume all the normal up-and-down angles encountered in the actual operation of a submarine. The instruments and controls are mounted in their relative locations and all function just as when actually installed on a submarine.

201

Figure 21-1. The diving trainer

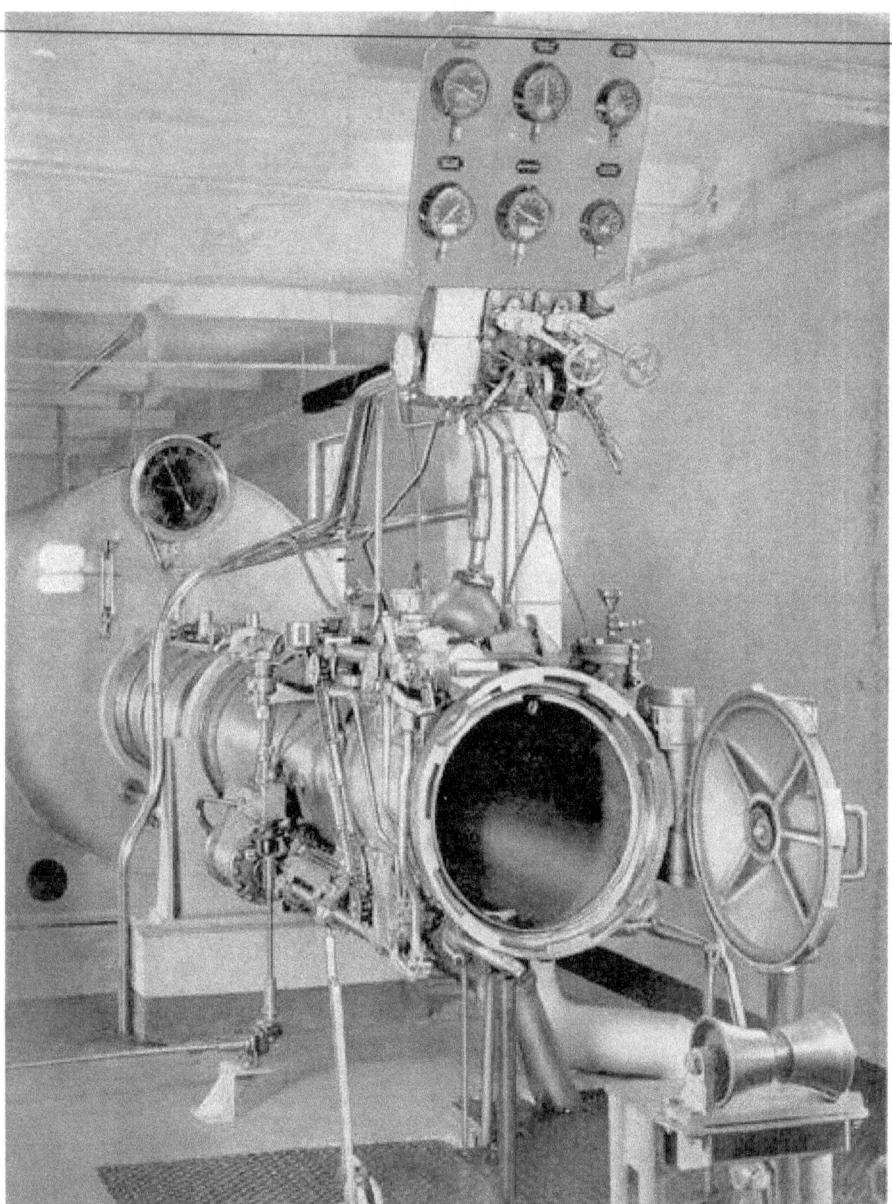

Figure 21-2. The torpedo tube trainer

The device is operated from a control stand in front of the room. This stand is the station of the instructor and affords him a full view of all the gages, instruments, and members of the control room personnel. The controls not normally installed on the officer, then issues the necessary orders to attain the desired condition of the submarine. Every action necessary to diving, trimming, and surfacing the ship is carried out to the most exact detail and the response of the tilting section of the control room and the registry of the port side of the control room are mounted in their relative positions at the side of the lecture room.

Electric and hydraulic controls enable the instructor to create conditions which are normally encountered and registered on the instruments. The student, acting as diving gages indicate to the student the same result as would be obtained with an actual submarine.

The use of this trainer has shortened the training period otherwise necessary and permits the training of an increased number of students.

D. THE TORPEDO TUBE TRAINER

21D1. Description. A recent innovation in training devices is an installation of a standard torpedo tube with all controls, interlocks, and mechanisms, and from which a standard torpedo may be fired and the procedure observed. (See Figure 21-2.)

The tube itself is an exact duplicate of those installed in the fleet-type submarines. The tube is mounted on the end of a water tight tank in which varying pressures, corresponding to assumed depths, can be attained. The torpedo tube muzzle extends into the tank and is fitted with the standard muzzle door. Guides along the bottom of this tank, and above the course of the fired torpedo, are installed to prevent any erratic course once the torpedo leaves the tube.

Heavy glass windows in the side of the tank permit observations to be made of the torpedo's exit from the tube and the amount of bubble checked.

The torpedo is loaded into the tube and is fired in a normal manner and at normal speeds. At the rear end of the tank the torpedo enters a restricted passage. The water expelled from this passage exhausts through gradually reduced orifices and the torpedo is brought to a gentle stop. It is then pushed back through the tank and tube to the loading rack where it may be used again.

A gyro angle indicator regulator is mounted near the tube and any normal operation connected with torpedo fire may be duplicated.